HIGH-PERFORMANCE LIQUID CHROMATOGRAPHY

HIGH-PERFORMANCE LIQUID CHROMATOGRAPHY

John H.Knox

John N.Done

Anthony F.Fell

Mary T.Gilbert

Andrew Pryde

Richard A.Wall

Edinburgh University Press

22 George Square, Edinburgh

ISBN 0 85224 341 3

Printed in Great Britain by
Brown, Knight & Truscott Ltd.,
Tonbridge, Kent

CONTENTS

PREFACE

High-Performance Liquid Chromatography (HPLC) is one of the most rapidly expanding techniques in analytical chemistry, with a doubling period of around two years. This book is the outcome of a pair of successful courses mounted by the Scottish Region of the Analytical Division of the Chemical Society in June 1977 in which the staff of the Wolfson Liquid Chromatography Unit of the Chemistry Department of Edinburgh University played a major role.

The course involved lectures, seminars and practicals. In the last of these we were fortunate in having the co-operation of a number of leading manufacturers of HPLC equipment who mounted their own experiments to demonstrate principles and applications of HPLC.

The book therefore combines theoretical treatment of the subject (arising from the lectures), discussion of equipment, its role, function and optimisation (arising from the seminars), and practical instruction on how to set up and run HPLC systems (arising from the practical experiments). The text does not claim to be a comprehensive review of the HPLC literature, for this has been done elsewhere, but it does claim to give the user, or intending user, of HPLC clear guidance as to how to operate and apply the technique in his own laboratory. We believe that the course team have been able to produce a text that is much more unified than is normally possible in a multi-author work since all have worked together in the same laboratory over a period of years, and hold roughly the same views about how HPLC can best be presented. The text is copiously illustrated with examples taken from our own work and from the literature. We sincerely hope that it will prove a valuable addition to the library of the practising liquid chromatographer.

The authors are jointly indebted to a variety of people: Mrs M.Duncan, secretary to the Wolfson Liquid Chromatography Unit, for typing the original draft manuscrupt; Mr D.J.L.Stewart-Robinson, of the Edinburgh Regional Computing Centre, for redrawing the diagrams; and Mrs J.MacDonald and Miss P.McManus, of the Centre for Tropical Veterinary Medicine, Easter Bush, Roslin, for preparing the camera-ready copy.

AFFILIATIONS

John H. Knox, Editor, Professor of Physical Chemistry, Director of Wolfson Liquid Chromatography Unit, Department of Chemistry, University of Edinburgh.

John N. Done, Research Chemist, Inveresk Research International, Inveresk.

Anthony F. Fell, Lecturer in Pharmacy, Heriot-Watt University, Edinburgh.

Mary T. Gilbert, Research Chemist, Wolfson Liquid Chromatography Unit, Department of Chemistry, University of Edinburgh.

Andrew Pryde, Research Chemist, Maag, A.G., Dielsdorf, Switzerland.

Richard A. Wall, Lecturer in Chemistry, University of Edinburgh.

1. INTRODUCTION

1.1 HISTORY AND GENERAL PRINCIPLES OF HPLC

High-performance liquid chromatography, HPLC, bears a close resemblance to gas chromatography, GC. It arose in the late 1960s out of experience in GC and from the application to LC of theories developed for GC. The major theoretical influence came from the work of Giddings (1), although Hamilton (2) had been a pioneer in his work on ion-exchange chromatography. The major experimental advances, which brought about practical HPLC systems, were made by Lipsky in 1967 (3), Huber in 1967 (4) and Kirkland in 1969 (5). HPLC complements GC, being able to separate substances that cannot readily be volatilised, and is particularly suitable for the separation of compounds having one or more of the following characteristics: (a) high polarity; (b) high molecular weight; (c) thermal instability; (d) a tendency to ionise in solution.

In HPLC, as in GC, the operating conditions, namely column temperature, inlet pressure, flow rate, etc., are closely controlled; the column is used repeatedly; the sample for analysis is small and is injected directly by valve or syringe onto the column; the separated solutes are detected as they emerge from the column by a sensitive detector; the detector signal is recorded to give a quantifiable record.

The main differences between HPLC and GC arise from the use of a liquid in place of a gaseous eluent. This has two major consequences:

(i) Because liquids are 20 to 100 times as viscous as gases, operating pressures are increased by this factor going from GC to HPLC.

(ii) Because rates of diffusion in liquids are 3000 to 30,000 times lower in liquids than in gases, particle dimensions in HPLC should in theory be around 50 to 200 times smaller (square root of the diffusion ratio) than for GC if equilibration times and hence analysis times are to be comparable. Although this reduction in scale is not quite achieved in practice, columns for HPLC are much smaller than in GC, being typically about 5 mm in bore and 100–250 mm in length. They are packed with 5–10 μm particles and operated with pressures in the range 20–200 bar (300–3000 psi).

Equipment for HPLC must be designed to accommodate these small columns without loss of performance and to withstand the high column inlet pressures. Accordingly, the high-pressure sections of the equipment must be made of metal. Since a wide range of fluids, from organic hydrocarbons to aqueous acids and alkalis, are used as eluents, metal components are made of grade 316 stainless steel and plastic components are preferably of PTFE or some other appropriately inert polymer. An important consequence arising from the small size of the columns is that the volumes of detectors, and that of all connecting tubing, must be minimal. Injection valves must be made to very high specifications if unswept dead volumes, which can cause serious peak distortion, are to be eliminated. Thus HPLC equipment demands a degree of precision micro-engineering not required in most GC equipment, and this has to be achieved in materials that are not the easiest to machine.

1.2 EQUIPMENT IN BRIEF

Although equipment is dealt with fully in chapters 8 to 10 the essential components are briefly discussed here. They are shown as a block diagram in figure 1.1.

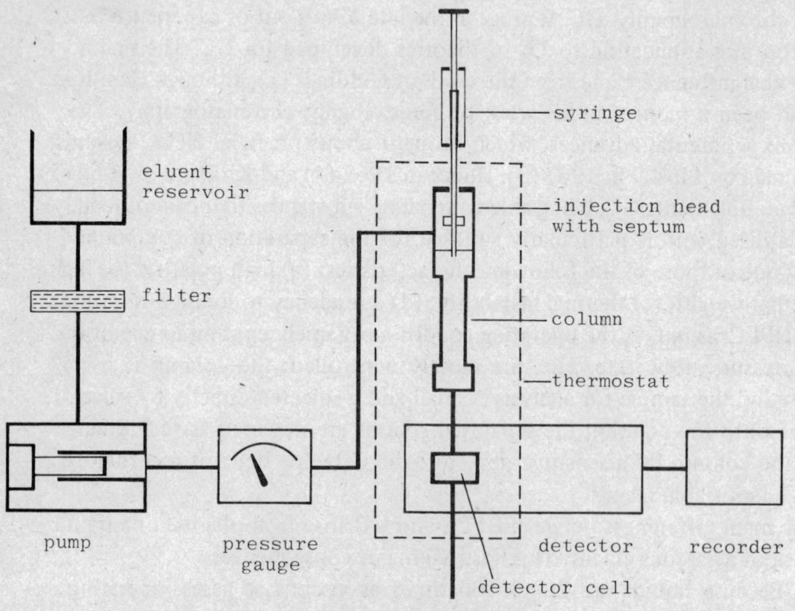

Figure 1.1. Essential components of a high-performance liquid chromatograph.

Eluents for HPLC may comprise water, aqueous buffer solutions, aqueous/organic mixtures, organic liquids or mixtures of organic liquids. The eluent viscosity is normally between 2×10^{-4} (pentane) and 2×10^{-3} N s m^{-2} (50:50 methanol/water). Freshly prepared eluents often contain undesirable dissolved air. This is generally removed by boiling under reflux or by evacuation before placing the eluent in the reservoir of the chromatograph, although a few commercial chromatographs allow degassing to be carried out in situ. Because nearly all pumps contain non-return valves whose function can be seriously impaired by dust particles, a fine mesh filter is placed between the reservoir and pump. The type and mesh of filter depends upon the type of pump, but the finest possible mesh that allows the pump to refill properly should be chosen. A typical filter will have a porosity of about 10 μm.

Pumps or eluent pressurisers are of five main types: (a) coil pumps in which inert gas at high pressure drives eluent out of a narrow tube; (b) air-driven pressure intensifiers (amplifiers) in which moderate air pressure drives a large-area piston connected rigidly to a much smaller-area piston bearing on the eluent; (c) reciprocating piston pumps of fixed frequency (usually between 0.5 and 2 Hz) and variable stroke; (d) large-volume syringe pumps driven by a lead screw and stepping motor; and (e) dual small-volume syringe or piston pumps driven by cams so designed to give constant flow at a pre-settable rate (variation of flow rate is obtained by changing the motor speed rather than the stroke). Types (a) and (b) provide a constant inlet pressure, while types (c), (d) and (e) provide a constant inlet flow. Both types are satisfactory, and there is no overriding argument in favour of either constant-flow or constant-pressure operation. The choice of pump very often depends upon the funding available.

In nearly all chromatographs the pressure of eluent is measured by a Bourdon pressure gauge or a strain-gauge pressure transducer. Flow-through pressure-measuring devices are desirable to avoid unswept dead volumes.

Columns vary considerably in dimensions, and have been as long as 1 m or even 3 m. However, current trends are towards short, fat columns, say 100 to 250 mm in length and 4 to 8 mm in bore. The column is best coupled directly to the injection unit, and this critical part in the design of any chromatograph is dealt with specifically in chapter 8. Typically, injection is carried out by micro-syringe, of 1, 5 or 10 mm^3 volume, the needle passing through an elastomer septum as in GC. Because of the high pressures used in HPLC the septum must be very well supported in the injector. Maximum pressures for injection by glass micro-syringes are about 100 bar (1500 psi). Ideally the sample should be placed sharply on the centre of the top of the packing material, but in practice is is usually injected into a 5 mm layer of glass beads, glass wool, or porous plastic, which tops the packing proper. This prevents disturbance of the top of the packed part of the column, which can seriously impair performance.

Eluate from the column passes through a minimum length of fine tubing, which must have a bore not larger than 0.25 mm (0.010 in), and into a detector cell. By far the most popular detector is the UV photometer, either of the fixed-wavelength type (254 nm Hg line generally) or of the variable-wavelength type. Modern UV detectors are stable to about 10^{-4} absorbance units and, with the usual 10 mm-long cell of 1 mm bore, are therefore capable of detecting down to 1 part in 10^9 by volume of a solute with a decadic molar extinction coefficient of 10^4 mol^{-1} dm^3 cm^{-1}.

Modern liquid chromatography can be carried out in any of the classical modes, of which the most important are liquid-solid adsorption chromatography, liquid-liquid partition chromatography, liquid-organobonded phase chromatography, ion-exchange chromatography and size-exclusion chromatography. These different LC modes are described in chapters 3 to 7.

Typically, a modern liquid chromatograph will give separations in 1 to 30 min and provide a recorder trace comprising a series of peaks such as that shown for a test mixture in figure 1.2. The figure illustrates the excellent peak sharpness and high speed at which separations can be achieved in HPLC under good conditions.

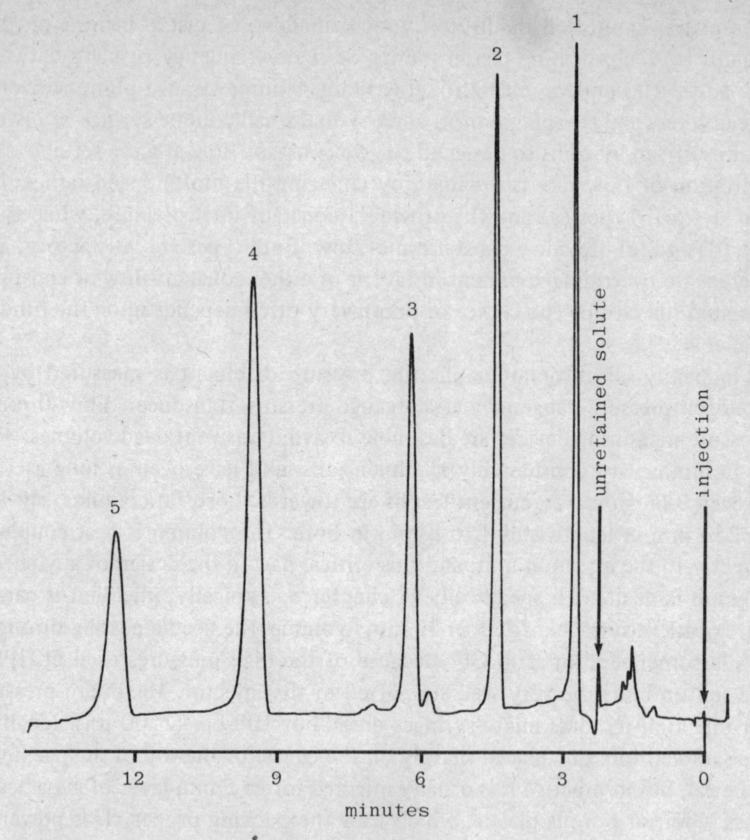

Figure 1.2. Test Chromatogram.

Column:	125 x 5 mm
Packing:	6 μm SAS Hypersil
Eluent:	methanol/water (40:60 v/v)
Solutes:	(1) acetone, (2) phenol, (3) p-cresol, (4) 3,5-xylenol, (5) phenetole
Pressure:	40 bar
Detector:	UV photometer, wavelength 254 nm, sensitivity 0.05 absorbance units full scale deflection (AUFS)

REFERENCES

1. Giddings, J.C., *Dynamics of Chromatography Part I.* Marcel Dekker, New York, 1965.
2. Hamilton, P.B., Bogue, D.C. and Anderson, R.A., *Anal. Chem. 32* (1960) 1782.
3. Horvath, C., Preiss, B. and Lipsky, S.R., *Anal. Chem. 39* (1967) 1422.
4. Huber, J.F.K. and Hulsman, J.A.R.J., *Anal. Chim. Acta 38* (1967) 305.
5. Kirkland, J.J., *J. Chromatogr. Sci.* 7 (1969) 7.

2. RETENTION AND PEAK SPREADING

The aim of column technology in HPLC may be defined as the achievement of the optimum combination of resolution of solutes, speed of elution, and economic use of pressure. The design of all other components of an HPL chromatograph must be governed by the most desirable configuration and operating conditions for the column.

The key to resolution in any form of chromatography is the proper combination of the differential migration of solutes and the control of band spreading. Fortunately, control of the migration rates, and of band spreading, can be treated almost independently, since the former is governed by thermodynamic (or equilibrium) considerations, while the latter is governed by kinetic considerations.

Before proceeding with the detailed arguments some basic terms must first be defined.

2.1 DEFINITIONS

The time that elapses between injection and elution of a solute is called the *retention time* or *elution time*, t_R, and the volume of eluent passed into the column during this time is called the *retention* or *elution volume* of the solute, V_R. These quantities are illustrated in figure 2.1. Elution time and volume are connected via the *volumetric flow rate*, f_v, i.e.

$$V_R = t_R f_v \tag{2.1}$$

If V_m is the volume of eluent in the column, the time to pass V_m into the column is denoted by t_m, and is equal to the retention time of a so-called unretained solute (i.e. one that behaves like eluent).

The *phase capacity ratio* k' is a measure of the degree of retention of a solute and is defined by

$$k' = \frac{t_R - t_m}{t_m} = \frac{V_R - V_m}{V_m} \tag{2.2}$$

The rate of movement of eluent along the column is called the *linear flow velocity* and is denoted by u; it is related to the retention time t_m and column length by

$$u = L/t_m \tag{2.3}$$

and to the volume flow rate by

$$u = f_v/a_m \tag{2.4}$$

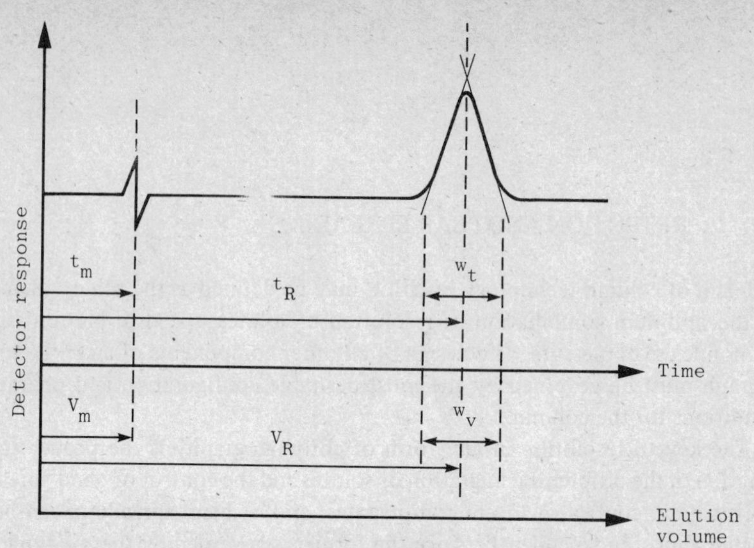

Figure 2.1. Parameters for defining retention and peak width.

where a_m is the mean cross-sectional area of the eluent phase within the column. The linear velocity of a solute band, u_{band}, is inversely proportional to its elution time, i.e.

$$u_{band}/u = t_m/t_R = 1/(1 + k') \tag{2.5}$$

The peak width at the base in the elution record (see figure 2.1) is denoted by w_t when measured in time units, in volume units by w_v, and as a distance within the column by w_z. Both theory and practice agree that the width of a band within the column w_z, increases as the square root of the distance migrated. Because of this it is convenient to define the band dispersion parameter, which is called the *height equivalent to a theoretical plate*, H, as

$$H = \frac{1}{16}\left(\frac{w_z^2}{L}\right) = \frac{L}{16}\left(\frac{w_t}{t_r}\right)^2 \tag{2.6}$$

The plate height can be thought of as the thickness of a transverse slice of column, and its dimensions are those of column length. The plate height concept was originated by Martin and Synge (1). The number of plates to which a column is equivalent is given by

$$N = L/H = 16\,(L/w_z)^2 \tag{2.7a}$$

$$= 16\,(t_R/w_t)^2 \tag{2.7b}$$

$$= 16\,(V_R/w_v)^2 \tag{2.7c}$$

The higher N, the better the resolution that can be expected from the column, and for a good column N will be between 1000 and 20,000.

Resolution of two compounds, A and B, is defined as the peak separation divided by the mean peak width, i.e.

$$R_s = \frac{t_{R(B)} - t_{R(A)}}{\frac{1}{2}(w_{t(B)} + w_{t(A)})} = \frac{V_{R(B)} - V_{R(A)}}{\frac{1}{2}(w_{v(B)}) + w_{v(A)})} \tag{2.8}$$

It is readily shown using the equations for N, t_R, w_t, k' that resolution can also be expressed (2,3) by

$$R_s = \frac{1}{2}\frac{(a-1)}{(a+1)}\left(\frac{\overline{k}'}{1+\overline{k}'}\right) N^{1/2} \tag{2.9}$$

where $a = (k'_B/k'_A)$ and $\overline{k}' = \frac{1}{2}(k'_A + k'_B)$. The three factors in the resolution equation emphasise three qualitatively obvious requirements for resolution: (a) solutes must be retained to different extents, i.e. $a \neq 1$; (b) solutes must be retained, i.e. $\overline{k}' \neq 0$; (c) the column must be equivalent to a minimum number of theoretical plates N.

2.2 SOLUTE RETENTION

Figure 2.2 is a formal representation of a packed chromatographic column. The essential feature of any chromatographic separation is the distribution of solutes to different extents between a flowing or mobile zone (the fluid outside the particles) and a fixed zone (the material, usually partly fluid, within the particles). Since it is essential for good plate efficiency that molecules equilibrate rapidly between the mobile and stationary zones, the distribution of any solute between zones is negligibly different from that which would be achieved at complete equilibrium (4). During chromatography, molecules move, while in the mobile zone, at the mean linear speed of the zone, u_o, and when in the stationary zone they do not, of course, move at all. It therefore follows (bearing in mind that equilibration is rapid and that any molecule is continually jumping from one zone to the other) that the speed of the band of solute molecules relative to the mobile zone is given by

$$u_{band}/u_o = \text{fraction of solute in mobile zone at equilibrium}$$

$$= \frac{q_{mz}}{q_{mz} + q_{sz}} = \frac{1}{1 + (q_{sz}/q_{mz})} = \frac{1}{(1 + k'')} \tag{2.10}$$

where q_{mz} and q_{sz} are the quantities of solute in the mobile and stationary zones respectively. The ratio q_{sz}/q_{mz}, denoted by k'' in equation 2.10, is called the *zone capacity ratio*.

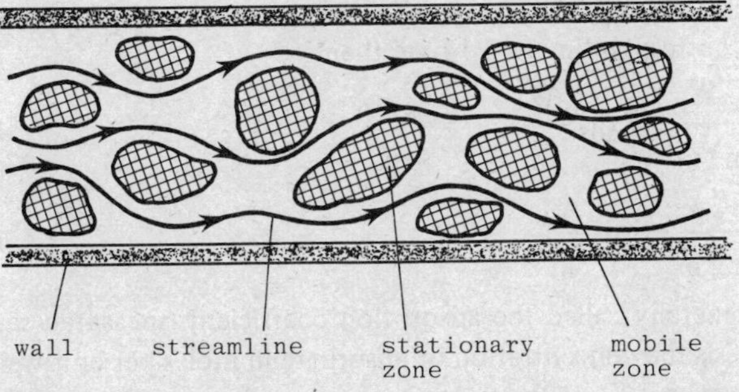

Figure 2.2. Formal representation of column structure showing stationary and mobile zones.

The mobile zone, being the fluid outside the particles, is not necessarily and indeed not generally identical to the eluent phase, since part of the eluent will be held within the pores of the packing material. Similarly the stationary zone, being the partitioning material inside the particles, does not generally correspond exactly to the stationary phase, which, depending upon the type of chromatography being performed, may be an adsorbent surface phase, a bonded layer, or a discrete liquid phase covering the internal surface of the support.

Only in exclusion chromatography is the rate of movement of a solute band referred to the rate of movement of the mobile zone. In all other forms it is referred to the mean rate of movement of the eluent phase, u. Making this change, equation 2.10 is modified to the form

$$\frac{u_{band}}{u} = \frac{q_m}{q_m + q_s} = \frac{1}{1 + q_s/q_m} \tag{2.11}$$

Comparing equations 2.11 and 2.5 immediately identifies k′ as

$$k' = q_s/q_m \tag{2.12}$$

Because of the near equilibrium assumption q_s and q_m can be regarded as equilibrium values and so can be given as

$$q_s = C_s V_s, \quad q_m = C_m V_m \tag{2.13}$$

where C_s, C_m are the equilibrium concentrations of the solute in the stationary and mobile phases, while V_s, V_m are the volumes of the stationary and eluent phases. Thus

$$k' = \frac{C_s V_s}{C_m V_m} = D \frac{V_s}{V_m} \tag{2.14}$$

where D is the equilibrium distribution coefficient for the solute between the two phases. Since the retention volume of a solute is inversely proportional to the band velocity we obtain, from equations 2.11 and 2.14

$$V_R = V_m(1 + k') = V_m + DV_s \tag{2.15}$$

When dealing with adsorption chromatography the "volume of the stationary phase" is not a particularly convenient concept, and the amount of stationary phase is more appropriately described by the weight of adsorbent in the column, W, or by the surface area of the adsorbent in the column, A. The appropriate equations for k′ corresponding to 2.14 are then

$$k' = \frac{C_s' W}{C_m V_m} = K_{ads} \frac{W}{V_m} \tag{2.16}$$

and

$$k' = \frac{C_s'' A}{C_m V_m} = K_a \frac{A}{V_m} \tag{2.17}$$

where K_{ads} is generally called the adsorption coefficient (measured say, in $cm^3 g^{-1}$) and C_s' is the concentration of adsorbate in moles per unit weight; or, alternatively, where K_a is the superficial adsorption coefficient (measured in $cm^3 m^{-2}$) and C_s'' is the adsorbate concentration in moles per unit area of adsorbent surface. The equations corresponding to 2.15 are then

$$V_R = V_m + K_{ads} W \tag{2.18}$$

$$V_R = V_m + K_a A \tag{2.19}$$

The variation of k′ with temperature (assuming that the phase ratio is independent of temperature) then depends upon the heat of transfer of the solute from the mobile to the stationary phase and follows the van't Hoff equation

$$\frac{d \ln k'}{dT} = \frac{\Delta H_{(m \to s)}}{RT^2} \tag{2.20}$$

Since $\Delta H_{m \to s}$ is much smaller in LC than in GC, temperature has much less effect on degree of retention in LC than in GC. Accordingly the great majority of LC separations are carried out at ambient temperature. When a column is thermostatted above ambient temperature this is usually done for one of the following reasons: (a) to allow easy thermostatting of the column in order to maintain k′ values constant; (b) to improve the equilibration rate between the zones; (c) to reduce eluent viscosity and improve separation speed.

In GC, retention and selectivity are controlled by adjusting the column temperature and stationary phase composition. In HPLC, change in the composition of the eluent serves both purposes, and therefore solvent programming (otherwise known as gradient elution) is used in HPLC where temperature programming would be employed in GC.

2.3 PEAK DISPERSION

Peaks should be made as narrow as possible relative to their time of elution to obtain the best resolution. That is (w_t/t_R) should be minimised (see figure 2.1). As shown by equation 2.7, this means maximising the number of plates to which the column is equivalent or minimising the plate height. To optimise performance this must be achieved within the pressure limitation set by one's equipment and without sacrificing analysis time. The problem of optimisation is therefore to achieve a desired plate count in the minimum time with the pressure available from the equipment (5).

A key equation in this connection is that for the pressure drop across the column, which can be written as

$$\Delta p = \frac{\phi \eta}{t_o} \left(\frac{L}{d_p} \right)^2 \tag{2.21}$$

where t_o is the elution time of a solute confined to the mobile zone (e.g. an excluded polymer), ϕ is a constant called the column resistance factor, whose value should be between 250 and 500, η is the eluent viscosity, L the column length and d_p the mean particle diameter. This equation is most often written in terms of the elution time of the eluent, t_m. It then takes the form

$$\Delta p = \frac{\phi' \eta}{t_m} \left(\frac{L}{d_p} \right)^2 \tag{2.22}$$

We then have $\phi'/\phi = t_m/t_o = V_m/V_o$. For most packing materials this ratio is between 1.8 and 2.2, so ϕ' is typically in the range 500 to 1000.

Either of these pressure-drop equations shows that for a given eluent (η fixed), packing structure (ϕ or ϕ' fixed) and elution time (t_o or t_m fixed) Δp will be independent of particle size provided that the length of the column, L, is kept proportional to d_p. In other words, Δp will be constant for any group of columns of the same *reduced length* where the reduced length, λ, is defined by

$$\lambda = L/d_p \tag{2.23}$$

Choosing the best operating conditions is now seen to imply choosing the particle size d_p that will produce the highest value of N for a given eluent (η), elution time (t_m) and reduced length (λ). Since we can write

$$N = L/H = \frac{(L/d_p)}{(H/d_p)} = \frac{\lambda}{(H/d_p)} \tag{2.24}$$

it is most convenient to discuss the plate height not in absolute terms but in terms of the number of particles corresponding to H. This quantity, h, is called the *reduced plate height*.

$$h = H/d_p \tag{2.25}$$

h is a measure of the degree of band dispersion produced by the packing considered in relation to the particle size. A band-dispersion process may be defined as any process that tends to separate a molecule of a solute from its neighbours (not to be confused with the separation of molecules of different solutes due to differences in their degrees of retention). There are three such processes in chromatography; these are illustrated diagramatically in figure 2.3 and summarised below:

(a) Dispersion due to the tortuous nature of the flow through a packed bed. This dispersion has a weak positive dependence upon flow velocity.

(b) Dispersion due to axial molecular diffusion. This depends upon the time the band resides in the column, and is most important at low flow velocities when the residence time is large.

(c) Dispersion due to slow equilibration between the mobile and stationary zones. This arises because molecules in the stationary zone tend to get left behind when the main band passes over, while molecules in the mobile zone move ahead. Dispersion due to slow equilibration increases directly with the flow velocity and the equilibration time.

Each dispersion process contributes independently to the total plate height and, considered on a particle scale, depends upon the balance between the flow of eluent over a particle and the diffusion of solute across a particle. The mathematical expression for the total plate height, showing the velocity dependence of the three terms corresponding to the three dispersion processes is

$$h = A\nu^{\frac{1}{3}} + \frac{B}{\nu} + C\nu \tag{2.26}$$

ν is called the *reduced linear velocity* of the eluent and is defined as

$$\nu = ud_p/D_m \tag{2.27}$$

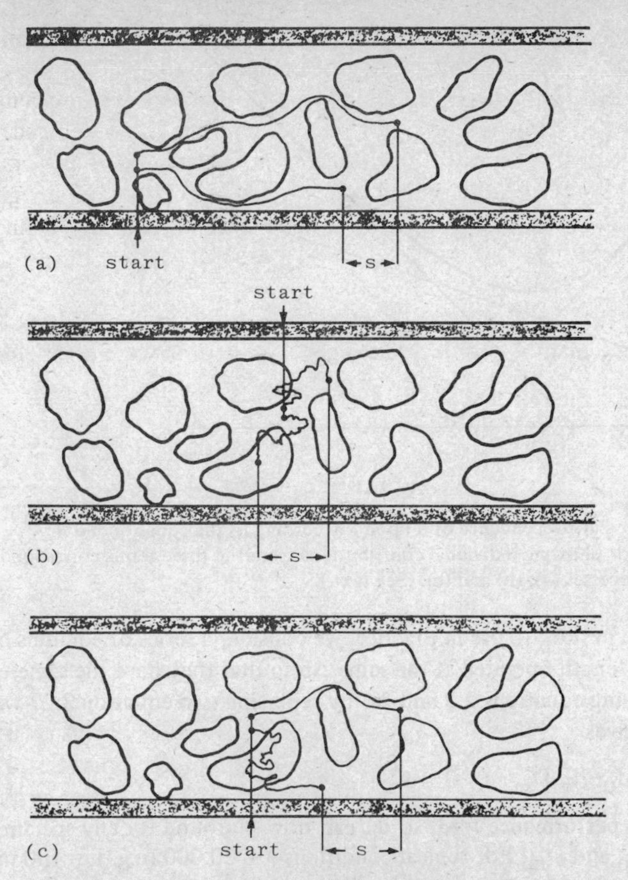

Figure 2.3. Major dispersion mechanisms in chromatography. ":" indicates two molecules initially adjacent; "s" represents their separation after operation of the dispersion process. (a) Dispersion by tortuous flow in mobile zone; (b) dispersion by axial diffusion (static conditions shown); (c) dispersion by slow equilibration between mobile and stationary zones.

The reduced velocity, a concept conceived by Giddings (4), is a measure of the rate of flow over a particle relative to the rate of diffusion within a particle. (Strictly the reduced velocity should be defined in terms of the linear velocity of the mobile zone, u_o, as $\nu_o = u_o d_p / D_m$, but, unfortunately, current practice always uses equation 2.27 and all published data have been reported on this basis.) The constants A, B and C are dimensionless, and for a good column have the approximate values $A < 1$, $B \approx 2$, $C < 0.1$.

Figure 2.4 plots equation 2.26, and includes lines representing the three terms of the equation. Since the dispersion caused by (b) is minimum at high velocities while those caused by (a) and (c) are minimum at low velocities, maximum efficiency corresponding to lowest h is achieved at an intermediate reduced velocity when the contribution from processes (b) and (c) are more or less equal. This velocity is, of course, that giving minimum h in figure 2.4, which occurs at a reduced velocity of $\nu \approx 3$.

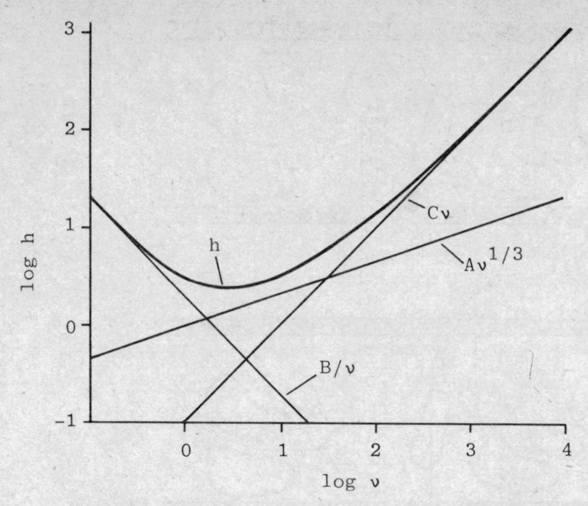

Figure 2.4. Dependence of h upon ν according to the equation $h = \nu^{1/3} + 2/\nu + 0.1\,\nu$ showing individual contributions from the three terms corresponding to processes (a), (b) and (c). (See text.)

To find the best d_p to use in practice, we consider a series of columns of the same reduced length operated at the same Δp so that they have the same t_m. We proceed using equations 2.3 and 2.5 by replacing u in equation 2.27 with $\lambda d_p/t_m$. This gives

$$\nu = \lambda d_p^2/t_m D_m \qquad (2.28)$$

For optimum performance $\nu \approx 3$, so d_p can now be found for any specified values of λ, t_m and D_m. For typical conditions $\lambda = 20{,}000$ (e.g. L = 100 mm, $d_p = 5\ \mu m$), $t_m = 100$ s, $D_m = 10^{-9}\ m^2\ s^{-1}$ (small molecule in water). With these values equation 2.28 gives

$$d_p = \left(\frac{\nu t_m D_m}{\lambda}\right)^{1/2} = \left(\frac{3 \times 100 \times 10^{-9}}{2 \times 10^4}\right)^{1/2}$$

$$\approx 4 \times 10^{-6} m = 4\ \mu m \qquad (2.29)$$

The pressure drop is obtained from equation 2.21 as

$$\Delta p = \frac{\phi' \eta \lambda^2}{t_m} = \frac{500 \times 10^{-3} \times (2 \times 10^4)^2}{100}$$

$$= 2 \times 10^7\ N\ m^{-2} = 20\ bar$$

For elution by water we deduce that particles of around 4 μm and quite low pressure drops of about 20 bar (300 psi) will be required. It should then be possible to obtain a plate count of about 8000 (i.e. h = 2.5). If one departs from the optimum conditions it is not difficult to show that one has to sacrifice either pressure drop, elution time or plate efficiency. For example, if 10 μm particles are used instead of 4 μm particles one can show (5) that the penalties incurred would be as follows:

(i) keeping N = 8000 and t_m = 100 s, then Δp = 8000 bar
(ii) keeping N = 8000 and Δp = 20 bar, then t_m = 250 s
(iii) keeping t_m = 100 s and Δp = 20 bar, then N = 4000

If one uses 20 μm particles it is impossible to obtain more than 2500 plates in 100 s whatever the pressure. If N and Δp were kept as in (ii) then t_m would be 720 s; and if t_m and Δp were kept as in (iii) then N would be 1600.

The effect of using too-large particles is not too severe with particles up to twice the optimum diameter, but departing from the optimum d_p by more than a factor of 2 imposes severe penalties. This was clearly demonstrated by Laird (6) in experiments where 6, 10 and 20μm spherical alumina was packed into columns of lengths calculated to produce 4000 plates with a pressure drop of 22 bar (330 psi). The results are shown in figure 2.5, and emphasise the superiority of the smallest particles. In the light of the above results there is no justification for using particles larger than 10μm for modern analytical HPLC. This does not, however, apply to preparative scale HPLC, where column overloading has a profound effect.

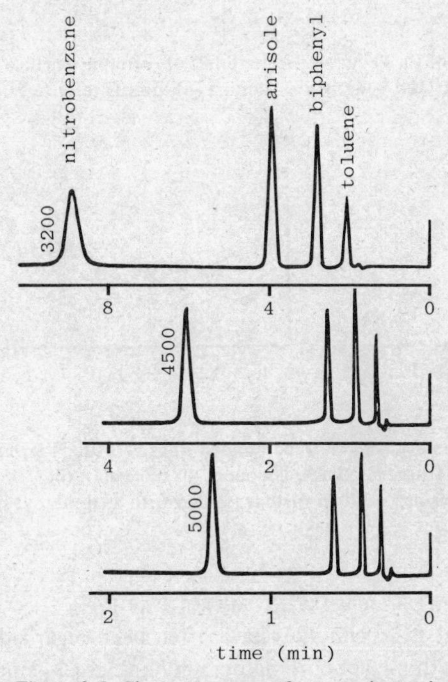

Figure 2.5. Chromatograms of a test mixture in columns containing Spherisorb Alumina of different particle size. Column lengths were calculated to give approximately 4000 theoretical plates with a pressure drop of 22 bar. (a) d_p = 20 μm, L = 500 mm, t_m = 105 s; (b) d_p = 10 μm, L = 135 mm, t_m = 35 s; (c) d_p = 6 μm, L = 90 mm, t_m = 16 s. The actual values of N are recorded on the final peaks.

The foregoing discussion of band dispersion has assumed that columns containing particles of different d_p do indeed give the same plot of reduced plate height h against reduced velocity ν. This view has excellent theoretical justification, but does it hold in practice? Over the last five years enough data have been accumulated to show that for well-packed columns containing particles from 5 μm to 50 μm there is indeed a common (h, ν) curve. This is demonstrated in figure 2.6 with data from references 6 and 7. In figure 2.7 similar data are shown for some chemically bonded materials (8), which are discussed in more detail below and in chapter 5.

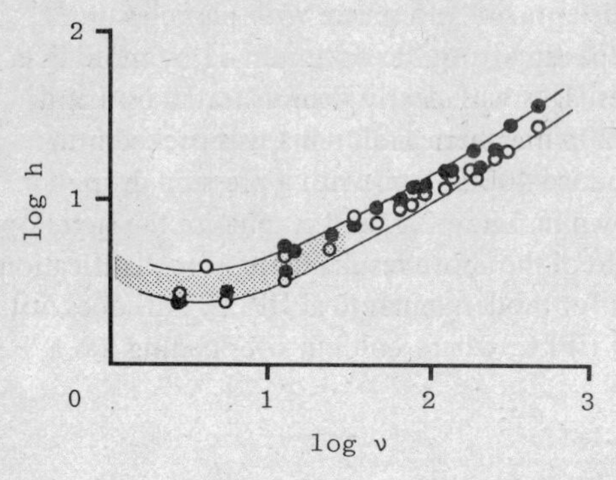

Figure 2.6. Superposition of (h, ν) curves for materials of different particle size. Shaded band refers to 6–10 μm Spherisorb Alumina (6), points refer to 50 μm Porasil (7).

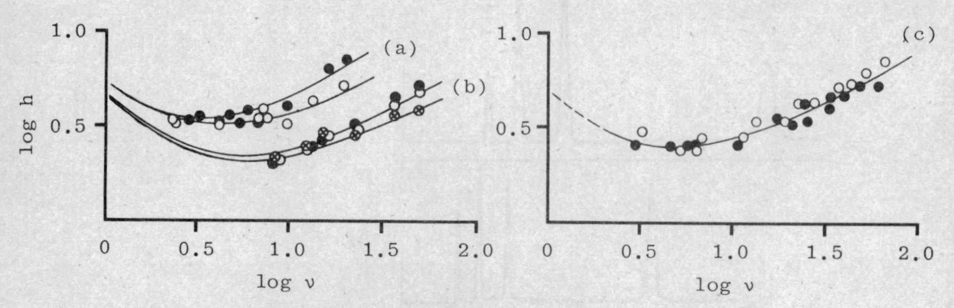

Figure 2.7. (h,ν) curves for silica gel and some bonded silica gels ($d_p \approx 9$ μm): (a) silica gel; (b) SAS silica (short alkyl chains bonded to silica gel); (c) ODS/TMS silica (octadecyl bonded to silica gel then further treated with a tri-methyl silyl reagent). (Data from ref. 8.)

2.4 THE DIFFERENT MODES OF LIQUID CHROMATOGRAPHY

In the past the various different LC techniques have often been regarded as separate and independent. Thus thin-layer chromatography has largely employed liquid-solid adsorption as the basis of retention. Paper chromatography has been the main form of liquid-liquid partition chromatography. Chromatography with ion-exchange resin beads has mainly been used for inorganic ions and for amino acids, and has used fairly elaborate equipment. Gel-permeation chromatography, with different equipment, has been regarded as independent of the other forms of LC, and has even been claimed not to be a form of chromatography at all.

The new equipment for HPLC, along with the improved understanding of LC that it has generated, has drawn all these different LC techniques together by demonstrating that all have a common theoretical basis and can be carried out with the same basic HPLC equipment.

The modes of liquid chromatography differ only in the nature, composition and structure of the stationary zone, and in the nature of the molecular forces that hold the solute molecules within the mobile and stationary zones.

The great majority of HPLC packing materials now in use are based upon wide-pore silica gels (i.e. silica gels with pores not less than 3 nm across). Such silica gels (9, 10) generally consist of fused aggregates of more or less spherical particles of colloidal silica, as shown diagramatically in figure 2.8. The colloidal particles are of the order of 2–20 nm in diameter and the surface area is from 100–400 $m^2 g^{-1}$. Silica gels with much wider pores, up to 100 nm diameter, can also be produced and these are finding increasing use in exclusion or gel-permeation chromatography (GPC).

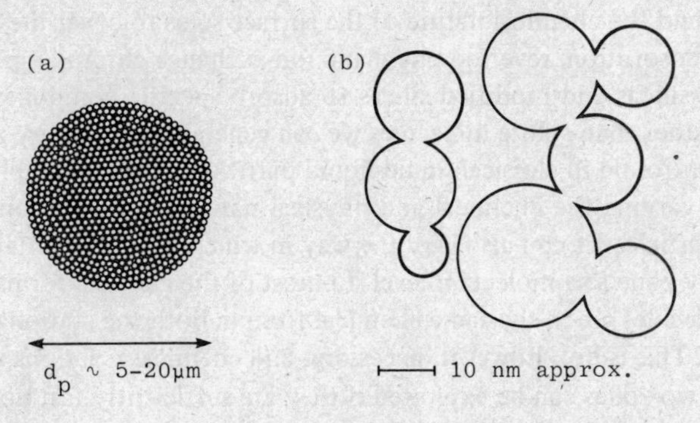

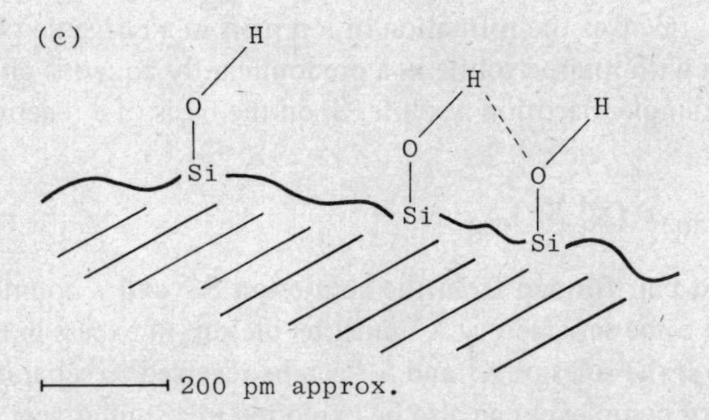

Figure 2.8. Structure of a microparticulate spherical silica gel, showing (a) the spherical particle composed of fused colloidal units; (b) the random array of fused colloidal units 5–15 nm in diameter; (c) the silanol groups at the silica surface.

The internal surface of a silica gel is covered with silanol groups (about four per nm^2) and these can be used to adsorb solute molecules in adsorption chromatography or to adsorb polar components (for example water, methanol, acetonitrile, acetic acid) from the eluent. In this way it is possible to form what amounts to a polar liquid layer over the internal surface. The silica gel then acts in effect as a support for a stationary liquid phase and so can be used for liquid-liquid partition chromatography.

Silanol groups may also be chemically reacted with compounds such as octadecyl trichlorosilane (8, 11) to give (after hydrolysis and further treatment) a substituted surface with hydrocarbon characteristics:

$$\geq Si\text{-}OH \longrightarrow \geq Si\text{-}O\text{-}Si(\text{-})\text{-}C_{18}H_{37}$$

These so-called bonded materials can be used for reversed-phase liquid chromatography where the eluent is, for example, aqueous methanol. Such hydrophobic surfaces, in contrast to the silanol surface of silica gel, will extract the less polar components from an aqueous eluent, so that what amounts to reversed-phase partition chromatography can be carried out. If instead of an alkyl group an amine or sulphonic acid is bonded to the surface, ion-exchange capability is introduced.

Thus it is seen that we can alter both the geometrical dimensions of the pores in silica gel and the chemical nature of the surface so as to cover the range of size-exclusion, adsorption, reversed-phase and ion-exchange chromatography. Further, by using silicas and modified silicas to adsorb specific components from the eluent rather than solute molecules we can generate a stationary zone that is close to that found in classical liquid-liquid partition chromatography.

In addition to varying the chemical and physical nature of the stationary zone within the particles we can also vary the way in which the solute interacts with the stationary zone at a molecular level. In most of the classical forms of LC the solute molecules are in the same chemical form in both the stationary and mobile zones. This is not, however, necessary, and chemical reactions within and between the two zones can be exploited if they are sufficiently fast not to act as a barrier to equilibration between the zones.

A form of distribution involving chemical reactions, which has recently sprung into prominence, involves the formation of ion pairs in an organic phase. These are in equilibrium with ionised solute in a predominantly aqueous phase (12–15). We can, for example, partition a solute, S, on the basis of a reaction such as

$$X^{+}{}_{(aq)} + S^{-}{}_{(aq)} \rightleftharpoons (X^{+},S^{-})_{org}$$

where (X^{+},S^{-}) is an ion pair formed from the solute ion S^{-} and a counter ion X^{+}. For the method to be satisfactory X^{+} must be present in excess in the aqueous phase. Evidently, the roles of X^{+} and S^{-} can be reversed to separate cations. The formation of complexes can also be exploited in a similar way if the equilibration is fast.

The selection of the optimum mode of operation for any HPLC application is still a fairly uncertain procedure, and there is no sure way to the method that will give the best results. Certain guidelines can, however, be laid out: the flow diagram in figure 2.9 presents one way of determining where to start, but when using any such scheme there is no substitute for experience. Thus while exclusion chromatography is the obvious method to use for chromatographing polymers one can obtain most valuable data by using adsorption chromatography, as shown by the chromatogram of polystyrene in figure 2.10.

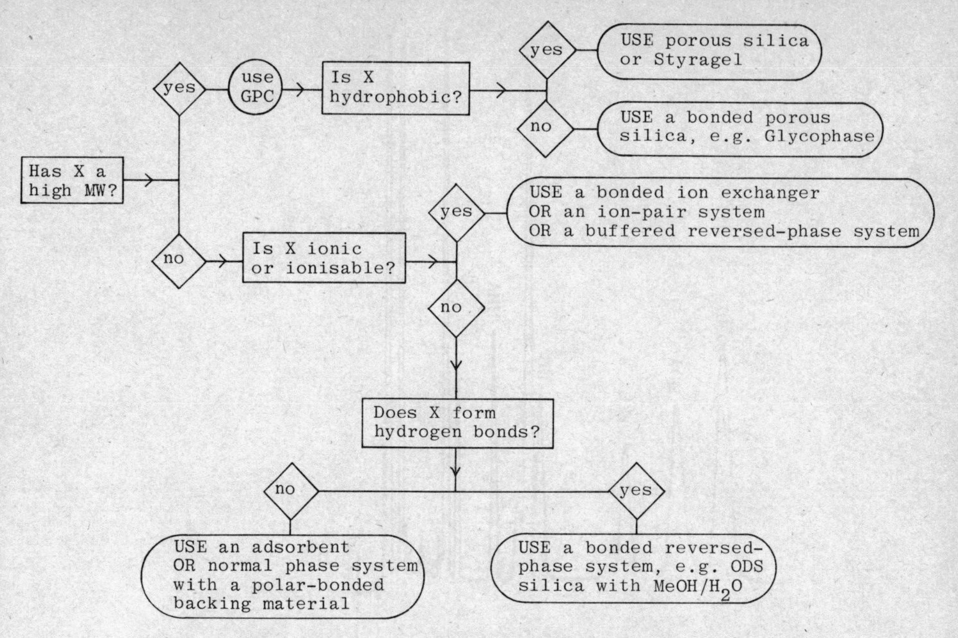

Figure 2.9. Flow chart for selection of the preferred LC mode on the basis of chemical type, functionality and molecular weight.

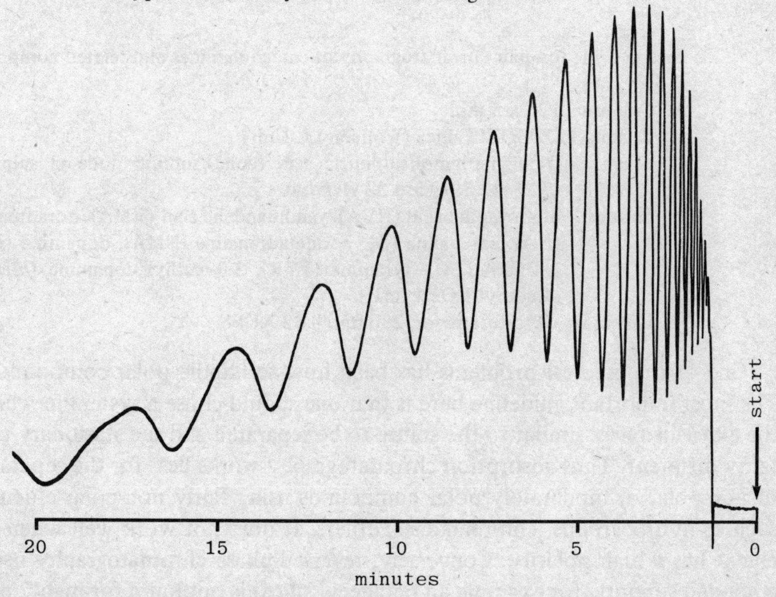

Figure 2.10. Adsorption chromatogram of a polystyrene standard on a small pore silica gel (F. McLennan).

Column: 100 x 7 mm
Packing: 6 μm Hypersil
Eluent: pentane/methylene chloride (75:25 v/v)
Solutes: polystyrene oligomers in a polystyrene standard M_n = 2000
Detector: UV photometer, 254 nm, 0.1 AUFS.

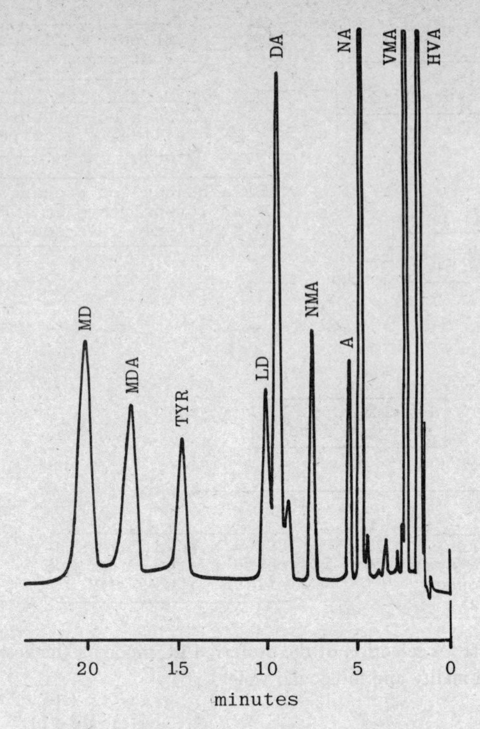

Figure 2.11. Ion-pair chromatography of catecholamines and related compounds (16).

Column: 125 x 5 mm
Packing: ODS/TMS silica (Wolfson LC Unit)
Eluent: water/methanol/sulphuric acid (conc)/sodium dodecyl sulphate (72.5:27.5:0.15:0.02 v/v/v/w)
Solutes: homovanilic acid (HVA), vanilmandelic acid (VMA), noradrenaline (NA), adrenaline (A), normetadrenaline (NMA), dopamine (DA), L-DOPA (LD), tyramine (TYR), 3-0-methyl dopamine (MDA), α-methyl DOPA (MD).
Detector: UV photometer, 280 nm, 0.02 AUFS.

One of the greatest problems has been how to handle polar compounds. The most important guideline here is that one should chose a system in which the eluent is fairly similar to the solute to be separated and the stationary phase fairly different. Thus adsorption chromatography works best for the separation of non-polar or moderately polar compounds using fairly non-polar eluents such as hydrocarbons, chloroalkanes, ethers. It does not work well when the eluent has a high polarity. Conversely, reversed-phase chromatography using a bonded support (for example an octadecyl silica) is optimum for highly polar solutes, which would be irreversibly retained on, say, silica gel. The eluents commonly used in such cases are aqueous solutions of lower alcohols or of acetonitrile. Extremely hydrophobic packing materials should not be used with pure water, since water cannot penetrate the pores because of its high surface tension. When compounds are ionic or ionisable it is natural to chose an ion-exchange sytem. However, recent experience is increasingly showing the great advantages in terms of flexibility and resolution of using reversed-phase ion-pair systems: an aqueous eluent is used, which contains a suitably hydrophobic ion that can form a lipophilic ion pair with solute ions. A recent example of ion-pair chromatography using a detergent as pairing agent is shown in figure 2.11.

In a similar vein, complexing metal ions along with a pairing ion can be used with dramatic effect for the separation of unionised solutes.

REFERENCES

1. Martin, A.J.P. and Synge, R.L.M., *Biochem. J. 35* (1941) 1358.
2. Simpson, C., *Practical High Performance Liquid Chromatography,* chaps 1 & 2. Heyden & Sons Ltd, London 1976.
3. Snyder, L.R. and Kirkland, J.J., *Introduction to Modern Liquid Chromatography.* John Wiley, New York, 1974.
4. Giddings, J.C., *Dynamics of Chromatography Part I,* Marcel Dekker, New York, 1965.
5. Knox, J.H., *J. Chromatogr. Sci. 15* (1977) 352.
6. Laird, G.R., Jurand, J. and Knox, J.H., *Proc. Soc. Anal. Chem. 11* (1974) 310.
7. Kennedy, G.J. and Knox, J.H., *J. Chromatogr. Sci. 10* (1972) 549.
8. Knox, J.H. and Pryde, A., *J. Chromatogr. 112* (1975) 171.
9. Snyder, L.R., *Principles of Adsorption Chromatography.* Marcel Dekker, New York, 1968.
10. Iler, R., *Colloid Chemistry of Silica and Silicates.* Cornell University Press, Ithaca N.Y., 1955.
11. Grushka, E., *Bonded Stationary Phases in Chromatography.* Ann Arbor Science Publishers Inc., Ann Arbor, 1974.
12. Eksborg, S., Lagerstrom, P-O, Modin, R. and Schill, G., *J. Chromatogr. 83* (1973) 99.
13. Kraak, J.C. and Huber, J.F.K., *J. Chromatogr. 102* (1974) 333.
14. Persson, B-A and Karger, B.L., *J. Chromatogr. Sci. 12* (1974) 521.
15. Knox, J.H. and Laird, G.R., *J. Chromatogr. 122* (1976) 17.
16. Knox, J.H. and Jurand, J., *J. Chromatogr. 125* (1976) 89.

3. ADSORPTION CHROMATOGRAPHY

Adsorption chromatography was the original chromatographic technique. Tswett (1) and Day (2) both used it around 1900 to separate plant pigments and petroleum respectively, and it was Tswett who coined the word chromatography because of the coloured bands he produced when he washed a concentrated plant extract down a column of adsorbent. Until the late 1960s adsorption chromatography in columns had changed little since those days, although in the form of thin-layer chromatography it became a widely used separation method. Nowadays adsorption chromatography in columns is one of the forms of liquid chromatography that has become a rapid efficient analytical method under the general heading of high-performance liquid chromatography (HPLC).

3.1 STRUCTURE OF ADSORBENT SURFACE

The basis of the separation achieved by adsorption chromatography is the selective adsorption of the components of the mixture onto active sites on the surface of an adsorbent. In order to achieve this, eluent molecules already adsorbed have to be displaced, and the system can be thought of as a reaction producing this displacement. The active sites on the surface are normally hydroxyl groups, either isolated or hydrogen-bonded as noted in chapter 2.

Silica is the most widely used adsorbent, and most of the investigations into the properties and the surface structure of adsorbents have been made with silica. However, the classical work of Snyder (3) involved both silica and alumina, and much of his data is for the latter adsorbent. Fortunately the surface structures of the two adsorbents are sufficiently similar that for our present purpose, except for minor details, we can treat them as one, and in what follows only silica gel will be discussed (see figure 3.1).

The active sites on silica, as already mentioned, are hydroxyl groups. These are isolated in crystalline silica, none being in a position to interact with its closest neighbour. In amorphous silica, such as that used for chromatography, the arrangement of the hydroxyls is not as ordered and interaction between neighbours can occur (4). In particular, hydrogen-bonded groups are present, the number of which depends directly on the adsorbent pore diameter. The surfaces of narrow-pore silicas are covered mainly by hydrogen-bonded hydroxyls, whilst the surfaces of wide-pore silicas are covered mainly by isolated hydroxyls. If amorphous silicas are heat-treated surface changes occur and siloxane groups are produced, as shown in figure 3.1. These groups are weak adsorbent sites, and as such they are not important chromatographically; however, they are important in the understanding of deactivation of silicas by heat treatment. Loss of water in this manner from wide-pore silicas involves two adjacent

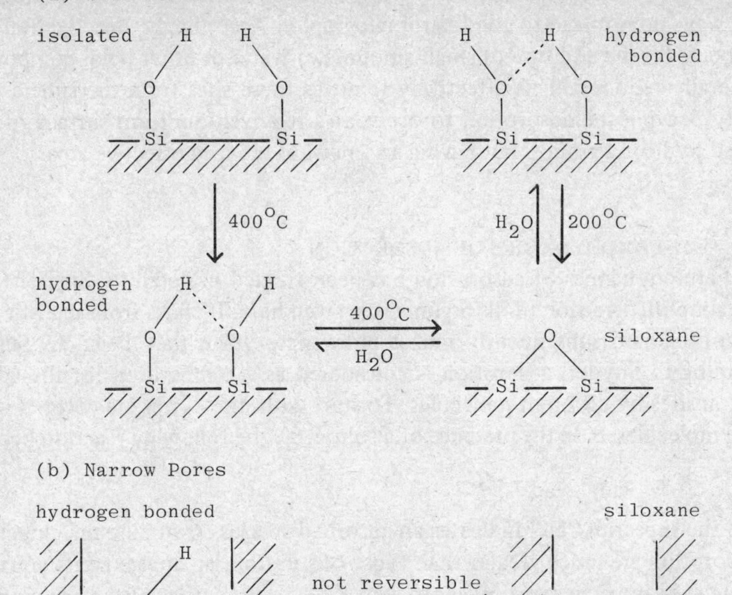

Figure 3.1. Surface processes occurring on heat and hydrothermal treatment of silica gel: (a) processes occuring in wide pores; (b) processes occuring in narrow pores.

hydroxyls, and the reaction can be reversed by heating in the presence of water. On the other hand, if the reaction occurs between hydroxyls on different but adjacent surfaces, such as those potentially present in small-pore silicas, no reverse reaction is possible and there is permanent loss of surface area. We can now understand the state of the surface of an amorphous silica prepared for HPLC. The surface is covered with hydroxyl groups, some isolated, others hydrogen-bonded, the numbers depending on the exact mode of preparation. If the material has not been heated above 100°C, many of these hydroxyls will be associated with water molecules, the majority of which can be removed by prolonged heating at 150–200°C. This is the normal activating procedure when using silica as a chromatographic adsorbent. Heating above 200°C alters the surface, producing siloxane groups from hydrogen-bonded groups. At still higher temperatures migration of individual groups occurs, resulting in further loss of hydroxyls. Depending on the pore diameter and the proximity of different surfaces, this last effect is accompanied to different extents by the permanent loss of surface area.

The two types of chromatographically useful active sites have different adsorption strengths. Hydrogen-bonded groups are considered to be significantly stronger sites for adsorption of anything but monofunctional solutes, and in many ways do not lead to good chromatography. Accordingly, deactivation of adsorbents by the addition of small amounts of water or other polar compounds is generally required. This effectively removes these sites from the sphere of activity, leaving the adsorption to occur at a relatively uniform surface of isolated hydroxyl groups, but having a somewhat reduced effective area.

3.2 THERMODYNAMICS OF ADSORPTION

The thermodynamics of adsorption has been treated in detail by Snyder (3), and a simplified version of his argument is given here. It starts from realistic assumptions and subsequently makes allowances when they begin to fail. According to Snyder, adsorption is considered as a competition for the adsorption sites by different molecules. To start with, these sites are covered with eluent molecules, S. In the presence of a solute, X, the following reaction occurs:

$$X + nS_{ad} \rightleftharpoons X_{ad} + nS \qquad (3.1)$$

where the subscript "ad" indicates an adsorbed species. Assuming the energies of adsorption are much greater than those of solution, an approximate equation relating the sample adsorption coefficient K_{ads} to properties of the adsorbent, the solute and the eluent can be derived (3):

$$\log_{10} K_{ads} = \log_{10} v_a + \beta(S^o - A_s\epsilon^o) \qquad (3.2)$$

K_{ads}, the adsorption coefficient, has been defined in chapter 2, and is the ratio of the moles of solute adsorbed per gram of adsorbent to the moles of solute per cm^3 of eluent. v_a is the volume of an adsorbed monolayer of solvent per gram of adsorbent. This monolayer is taken to have a thickness of about 0.35 nm. v_a is reduced when water or some other polar material is used to deactivate the adsorbent. Essentially v_a measures the specific area of the adsorbent available to the solute, which is assumed not to be able to displace any molecules of the deactivator. β is termed the surface activity of the adsorbent, and is arbitrarily given the value of unity for a thermally activated adsorbent that has not been deactivated. For adsorbents that have been sufficiently deactivated to be useful, β is in the range 0.5 to 0.9. v_a and β are properties of the adsorbent and are the same for all solutes and eluents.

The factor in brackets, $(S^o - A_s\epsilon^o)$, contains the properties of the eluent and solute. S^o is a dimensionless free energy of adsorption of the solute onto the surface (actually $S^o = \Delta G^o/2.303RT$), and is somewhat misleadingly termed the "energy of adsorption" by Snyder. A_s is the area of the solute molecule in units of 0.085 nm^2 (which is 1/6th of the area of an adsorbed benzene molecule and so corresponds to the effective area of an aromatic C atom on the surface), while ϵ^o is the hypothetical "energy of adsorption" of eluent covering 0.085 nm^2. Thus $A_s\epsilon^o$ is the "energy of adsorption" of the quantity of eluent required to cover the same area as the solute molecule. By convention ϵ^o is taken as zero for pentane. Values of ϵ^o for other solvents are then positive (with the possible exception of some fluorocarbons) and conveniently fall in the range

0 to 1. This implies that the true adsorption free energies of solvents at 25°C onto an area of 0.085 nm^2 exceed that for pentane by up to 6000 J mol^{-1}. The term in brackets in equation 3.2 is now seen to represent the "energy change" for reaction 3.1, that is for the replacement of adsorbed eluent by solute on a surface of unit activity ($\beta = 1.00$).

A more convenient form of equation 3.2 can be derived by noting (from equation 2.16) that

$$k' = K_{ads} \frac{W}{V_m} \tag{3.3}$$

where W = weight of adsorbent in column and V_m = volume of eluent in the column. This substitution for K_{ads} gives

$$\log_{10} k' = \log_{10}\left(\frac{v_a W}{V_m}\right) + \beta(S^o - A_s\epsilon^o) \tag{3.4}$$

$v_a W$ is the volume of solvent forming a monolayer coverage on the total active adsorbent surface within the column, and so can formally be equated to V_s, the volume of "stationary phase" in the column. We thus obtain

$$\log_{10} k' = \log_{10}\left(\frac{V_s}{V_m}\right) + \beta(S^o - A_s\epsilon^o) \tag{3.5}$$

Comparison with equation 2.14 shows that the distribution coefficient between eluent and stationary or surface phase is given by

$$\log_{10} D = \beta(S^o - A_s\epsilon^o) \tag{3.6}$$

Equation 3.4 gives an excellent description of the behaviour of simple solutes in adsorption systems using non-polar eluents. Under these conditions the basic assumptions are obeyed. The most important features of this equation are the natural emergence of an eluotropic series, i.e. an eluent strength series based on ϵ^o, and of a procedure,using simple compounds, that provides a means of standardising an adsorption system. Adsorption standardisation should always be carried out, where possible, since the calculation of k′ for standard compounds (i.e. with known S^o and A_s) with an eluent of known ϵ^o can lead to an estimation of log v_a and β.

The key solvent strength parameter ϵ^o, which relates to the eluotropic series, was calculated by Snyder for a wide variety of pure eluents and some of his values for alumina are shown in table 3.1. A full table is given in the appendix.

Snyder also extended his treatment to eluents composed of two components, and derived equations that can be used to calculate the ϵ^o values of such mixtures. Figure 3.2 shows the calculated and experimental values of ϵ^o for various eluent mixtures (5). Mixed eluents are used in two ways to achieve or improve separation. Firstly, intermediate values of ϵ^o can be produced as required, and, secondly, the differing solution energies that were neglected in the derivation of equation 3.2 can be brought into play in order to optimise resolution. The procedure involves the preparation of binary mixtures of different eluents so that they have the same ϵ^o values. In many cases it will be found that one of the

Table 3.1. Eluotropic series for alumina (3)

Eluent	ϵ^0	UV cut-off (nm)
Pentane	0.00	210
1-pentene	0.08	215
Carbon tetrachloride	0.18	265
2-chloro-2-methyl propane	0.30[a]	225
Methylene chloride	0.42	235
1,2-dichloroethane	0.49	225
Dioxan	0.56	220
Ethyl acetate	0.58	260
Diethylamine	0.63	225
Acetonitrile	0.65	190
Methanol	0.95	205

[a] Assumed the same as for 2-chloropropane

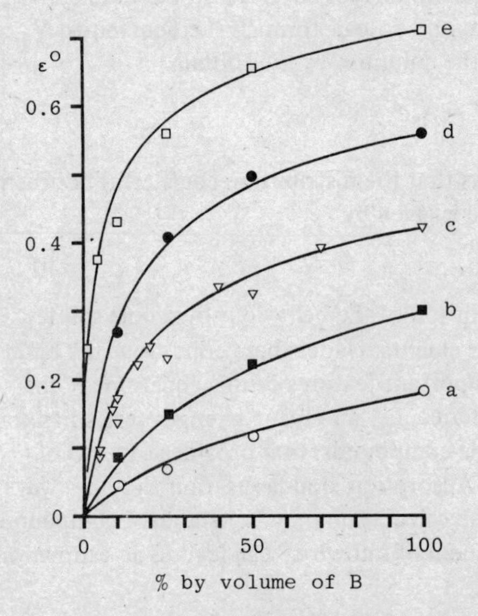

Figure 3.2. Dependence of ϵ^0 on composition of eluent for alumina with different solvents, B, added to pentane (3): (a) carbon tetrachloride; (b) propylchloride; (c) methylene chloride; (d) acetone; (e) pyridine. Points are experimental data, lines calculated.

mixtures will give the desired separation. Table 3.2 illustrates this process in detail.

This fairly simple theory works reasonably well and serves as a good basis for developing separations, but it begins to break down when the basic assumptions are no longer valid. This is particularly so when strongly polar eluents are required. The basic approach can, however, still be used qualitatively with suitable modification and, using this, much of the available data can be rationalised.

Table 3.2. Capacity factors for 1-acetoxynaphthalene and 1,5-dinitronaphthalene on alumina with different eluent mixtures, all having an approximate ϵ^o value of 0.25.

Eluent (volume % in pentane)	k_1' 1-acetoxy-naphthalene	k_2' 1,5-dinitro-naphthalene	$\alpha = \frac{k_2'}{k_1'}$
50% benzene	5.1	2.5	0.49
23% methylene chloride	5.5	5.8	1.05
4% ethyl acetate	2.9	5.4	1.9
5% pyridine	2.3	5.4	2.3
0.05% dimethylsulphoxide	1.0	3.5	3.5

3.3 CONDITIONING OF ADSORBENTS

The first important requirement for good reproducible adsorption chromatography is strict control of β, the adsorbent activity. This is accomplished by the addition of water or other polar material to eluent or adsorbent. Both β and v_a are thus reduced. The way these parameters depend on the amount of modifier has been studied in detail by Snyder (3). Figure 3.3 shows the effect of the amount of water on β, v_a and K_{ads} for a silica gel previously activated at 195°C for 4 hours. The control of the activity of the adsorbent by addition of water is easy when using classical "dry"-packed columns but the advent of "slurry"-packed columns in the last decade has made this method of control impossible. Nowadays the control is provided by adjusting the composition of the mobile phase and producing a dynamic equilibrium within the column between it and the adsorbent. With water as modifier this process is slow. This is largely because the solubility of water in non-polar organic liquids is very low, as shown by the data in table 3.3. In practice equilibration with water requires a fairly elaborate procedure. First it is necessary to decide upon the degree of water saturation required and to make up the eluent accordingly. For solvents like pentane, hexane, and the chloralkanes, the solvent is first passed through a column of

Table 3.3. Solubility of water (% w/w) in various organic liquids at 25°C from *Techniques of Organic Chemistry*, vol. VII (ed. Weissberger), Interscience Publishers Inc., New York, 1955).

Hexane, pentane	0.012
Carbon tetrachloride	0.010
Butyl chloride	0.08
1,2-dichloroethane	0.15
Methylene chloride	0.20
Diethyl ether	1.47
Ethyl acetate	3.3
Butan-1-ol	20.5

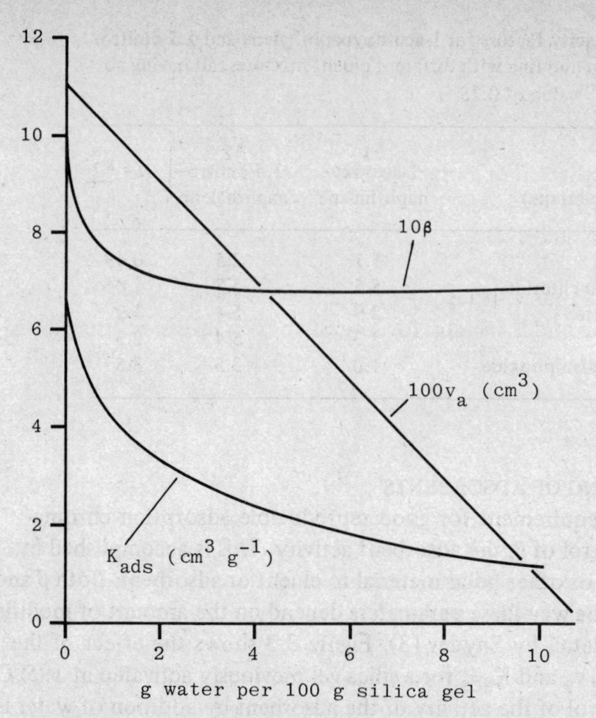

Figure 3.3. Dependence upon water content of surface activity β, surface phase volume, v_a, and adsorption coefficient, K_{ads}, of naphthalene dissolved in pentane. Data for a wide-pore silica gel of surface area 310 m^2 g^{-1} activated at 195°C for 4 h (3).

activated silica gel or alumina, or a Linde molecular sieve (typically a 500 × 20 mm tube containing 200 μm adsorbent). This removes all water and also serves to take out UV-absorbing materials and any polar impurities. A proportion of the purified and dried solvent is then passed through a similar column of water-saturated silica gel. This produces 100% saturated solvent. The "dry" and "wet" solvents are then mixed in appropriate proportions to give the partially saturated solvent desired. 50% saturation would very often be used.

Column equilibration must now be carried out. This can be achieved by passage of the part-saturated solvent, but the process can be very lengthy as shown by the following example. A fully activated silica gel of 200 m^2 g^{-1} surface area will typically require 70 mg of water per gram for full saturation. Fully saturated pentane at 25°C contains 0.12 mg of water per cm^3. Thus about 600 cm^3 of eluent will be required to saturate 1 gm of silica. This would represent about 700 column volumes of eluent. Such large volumes for conditioning a column are impracticable and wasteful. In practice equilibration is carried out using a more polar solvent, such as ether, which has the same degree of saturation as the desired eluent. Equilibration with respect to water is then achieved by passage of, say, 10 column volumes, and the ether is then replaced by the part-saturated solvent to be used as eluent. But the process is still tedious and time-consuming. Much more rapid equilibration is achieved by using polar organic modifiers in place of water. Methanol (ϵ^o = 0.95) and acetonitrile (ϵ^o = 0.65) are particularly suitable. Concentrations range from 0.02 to 0.2%. The use of an organic modifier is, however, only acceptable if the solutes to be separated are less strongly adsorbed. A detailed discussion of the whole subject has been given by Engelhardt (7). Variation of surface activity can also be achieved by heating silica to temperatures above 400°C (8). This treatment has to be carefully controlled, because irreversible changes take place and reactivation of the surface is slow and difficult.

Before packing a column with adsorbent a standard activation procedure should be carried out. Heating under vacuum for several hours is recommended. Silica can be heated to about 200°C whilst alumina is not damaged by temperatures up to 400°C. This treatment will yield reasonably active adsorbent with little or no water present, which can be easily packed using a slurry-packing procedure (see chapter 12) and conditioned as outlined above to give the required activity.

Another aspect of adsorption chromatography, described by Snyder as the "general elution problem", is closely related to the control of activity. Many samples will consist of compounds that are adsorbed to widely different extents and that cannot be eluted by a single eluent within a reasonable range of k′. In order to analyse such mixtures satisfactorily the eluent composition must be changed as the chromatography proceeds. This procedure is known as gradient elution, and is used in HPLC when column equilibration is rapid. This is the key factor, and in adsorption chromatography it is difficult to achieve. Various regeneration procedures have been published (9, 10), but for routine analytical use it would appear that gradient elution is best avoided in adsorption chromatography. It is nevertheless a very useful technique for investigating unknown samples. Equipment for gradient elution is described in chapter 9.

A final point concerning the use of silica (or alumina) as a column packing ought to be mentioned. With non-polar eluents the adsorption model developed by Snyder appears to give a good theoretical background to the experimental data. These systems are also genuine adsorption systems. However, with non-polar eluents modified with a reasonable amount (~ 5%) of polar solvent, a case can be argued for the systems no longer being an adsorption one but a partition one. In these cases, at equilibrium, a layer of the polar component is very likely to be built up next to the silica surface. Hence both adsorption and partition may be contributing to the separation mechanism. This difference may well be one of semantics, but it is worth mentioning to emphasise the fact that little can be gained by arguing the pros and cons of partition versus adsorption. After all, on practical grounds, it is not the mechanism but the fact of the separation that is important.

3.4 ADSORBENT CAPACITY AND PREPARATIVE CHROMATOGRAPHY

When the size of the sample applied to the column is gradually increased a stage is reached when the peak width increases, and the k′-value changes. The *sample capacity* or *load capacity* of a particular material is a measure of the material's resistance to these effects. Original work suggested that the loading at which peak broadening and change of k′ occurred were closely related. Recent work

(11) has shown this conclusion to be simplistic and that, with columns of very high efficiency, H is a linear function of the sample load at loadings of as little as 1 μg g^{-1}. Some typical plots are shown in figure 3.4. The reciprocal of the gradient of such plots seems to give the most realistic measure of adsorbent capacity, and is termed the *relative capacity* for the particular adsorbent. The relative capacity appears to depend on many of the chromatographic variables such as adsorbent activity, adsorbent surface area and the capacity factor k' of the solute. Figure 3.4 shows how H varies as the load is increased for a commercially available microparticulate silica and, in particular, illustrates the strong dependence on k'. The fact that, at low-enough loading, H is independent of k' is interesting, since it points the way to the conditions under which adsorption chromatography columns should be tested after they have been packed. This is discussed in detail in chapter 12.

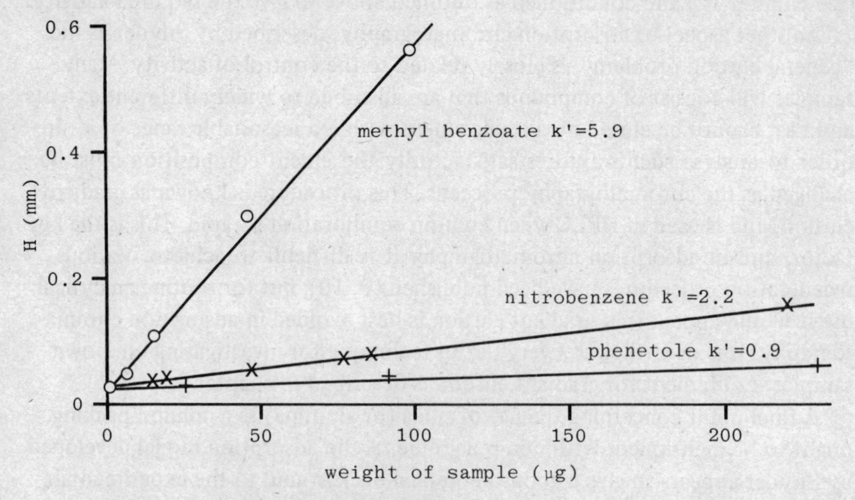

Figure 3.4. Effect of weight of sample on height equivalent to a theoretical plate, H, for a typical adsorbent (11).
Column: 145 x 5 mm
Packing: 10 μm Partisil (silica gel)
Column contained ~ 2 gm of packing material.

The effectiveness of an adsorbent (or indeed of any packing material) in preparative liquid chromatography is directly related to its load capacity, since the column will nearly always be operated in the overloaded condition, that is when the peaks are substantially wider than would be expected using analytical samples. Much work has been carried out on preparative aspects of adsorption chromatography and it now seems that the expertise is available to carry out quite large-scale separations in relatively short times. Various workers have discussed the relevant parameters, in particular the best particle size to use (11), the effect of the volume of sample (12), and the effect of flow rate (11). The general conclusions appear to be that, with reasonable selectivity, throughputs of up to a few hundred grams a day are now possible using the recently developed technology (13).

3.5 PACKING MATERIALS FOR ADSORPTION HPLC AND SELECTED APPLICATIONS

Table 3.4 lists the main adsorbents currently available for HPLC. Only microparticulate materials are included, because pellicular adsorbents are, we believe, now obsolete and mainly of historical interest. The 5 and 10 μm particles are suitable for analytical scale HPLC, while the 20 μm particles would be appropriate for use in preparative scale applications.

Table 3.4. Microparticulate adsorbents for high-performance liquid chromatography.

Name	Particle size (μm)	Surface area (m^2g^{-1})	Shape[a]	Supplier
Silica				
CPG	5–10	40–190	I	Pierce and Warriner
Hypersil	5–7	200	S	Shandon Southern
LiChrosorb Si	5,10,20	400	I	Merck
Nucleosil	5,7.5,10	500	S	Machery-Nagel
Partisil	5,10,20	400	I	Reeve Angel
μ-Porasil	10	350	S	Waters Associates
Sil-60	5,10,15,20	500	I	Machery-Nagel
Sil-X-I	13	400	I	Perkin Elmer
Spherisorb S	5,10,20	200	S	Phase Separations
Vydac 101-TP	10			Separations Group
Zorbax-Sil	6–8	300	S	Du Pont
Alumina				
Alox 60-D	5–20	60	I	Machery-Nagel
LiChrosorb Alox-T	5,10	70–90	I	Merck
Spherisorb A	5,10,20	95,175	S	Phase Separations

[a]S = spherical, I = irregular

The main differences between the listed materials are in their surface areas and particle shape. To a first approximation the capacity ratios and load capacities of the different materials will be proportional to their surface areas, other factors being kept constant. Particle shape appears to have little effect on column performance, if this is measured in terms of plate height for a given particle size whether the particles are large (14) or small (15). It is, however, generally believed that the column resistance parameter, ϕ', is higher for irregular particles than for spheres, but this has also been questioned (15). The problem in making such a comparison is to define clearly what is meant by the "particle size" and to establish the correlation of ϕ' with the range of distribution of particle size. In making a final decision on which material to use factors such as price, availability and batch reproducibility may be more important criteria.

Applications of high-performance adsorption chromatography are widely documented. In general the adsorption technique is used for the separation of non-polar or moderately polar materials. Table 3.5 lists a few of the more recent applications, while figures 3.5 to 3.8 illustrate the excellent efficiency that can be attained using modern materials and equipment.

Table 3.5. Selected applications of high-performance adsorption chromatography.

Sample	Adsorbent	Reference
Aflatoxins	silica	16
Optical brighteners	silica	17
Porphyrins	silica	18
Insecticides	silica	19
Phenothiazine-5,5-dixoides	alumina	20
Flavanoids	silica	21
Antibiotics	silica	22
Phospholes	alumina	23
Organometallic isomers	silica	24
Urea herbicides	silica	25
Plant protectants	silica	26

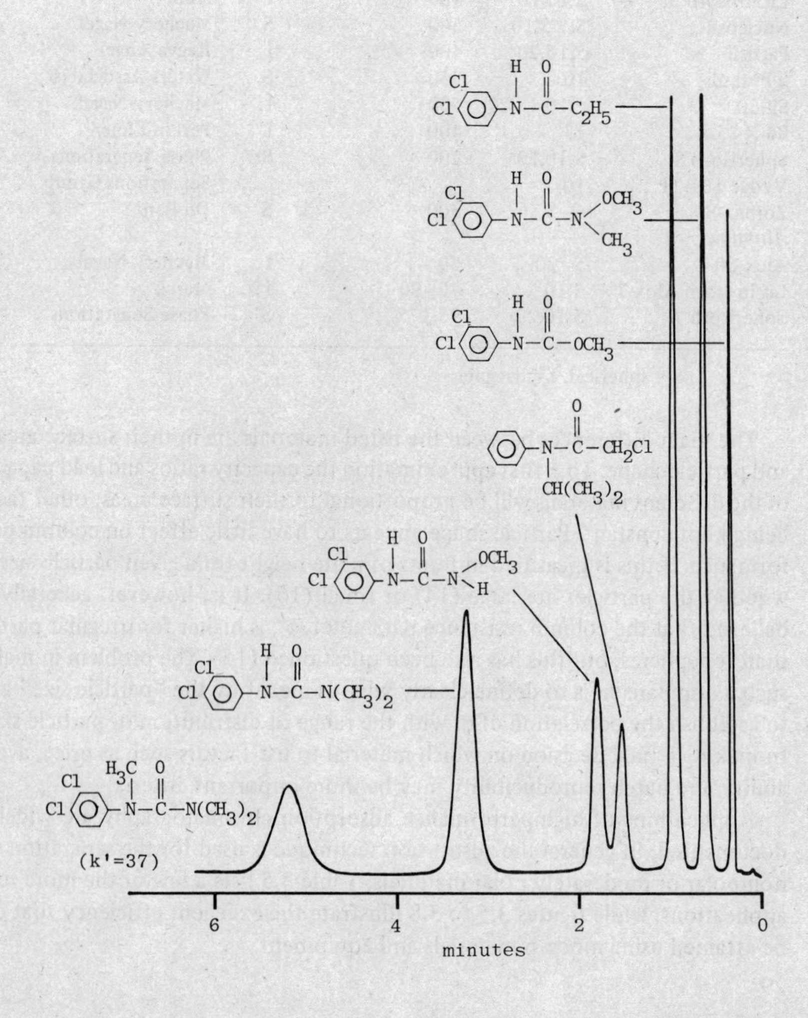

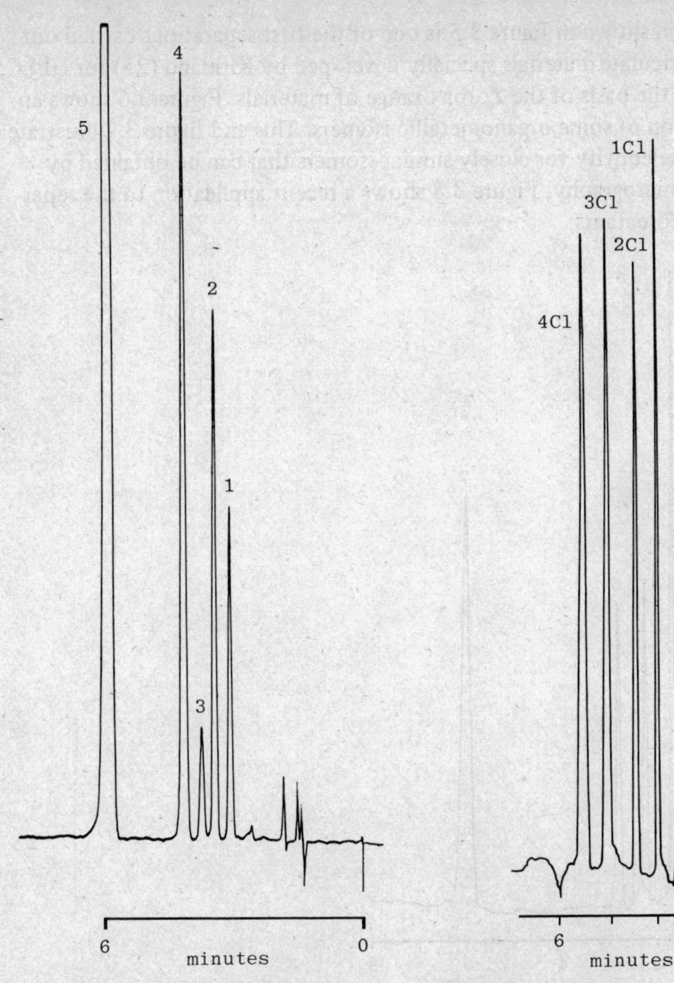

Figure 3.6. Adsorption chromatogram of five isomers of tricarbonyl (phenyl cycloheptatriene) iron (24).

Column: 125 x 5 mm
Packing: 6 μm Hypersil
Eluent: hexane
Detector: UV photometer, 254 nm, 0.1 AUFS

Figure 3.7. Adsorption chromatogram of four isomers of chlorophenothiazine-5,5-dioxide.

Column: 115 x 4 mm
Packing: 7.5 μm Spherisorb Alumina Y
Eluent: hexane/dioxan (50:50 v/v)
Detector: UV photometer, 254 nm

Figure 3.5. Adsorption chromatogram of substituted ureas and carbamates on microparticulate silica (25).

Column: 250 x 3.2 mm
Packing: 8–9 μm porous silica microspheres, surface area 250 $m^2\ g^{-1}$
Eluent: methylene chloride (50% water saturated)
Pressure: 135 bar
Detector: UV photometer, 254 nm, 0.1 AUFS

The separation shown in figure 3.5 is one of the first separations carried out on the microparticulate materials specially developed by Kirkland (25) for HPLC, and which form the basis of the Zorbax range of materials. Figure 3.6 shows an unusual separation of some organometallic isomers. This and figure 3.7 illustrate the remarkable selectivity for closely similar isomers that can be obtained by adsorption chromatography. Figure 3.8 shows a recent application to the separation of plant protectants.

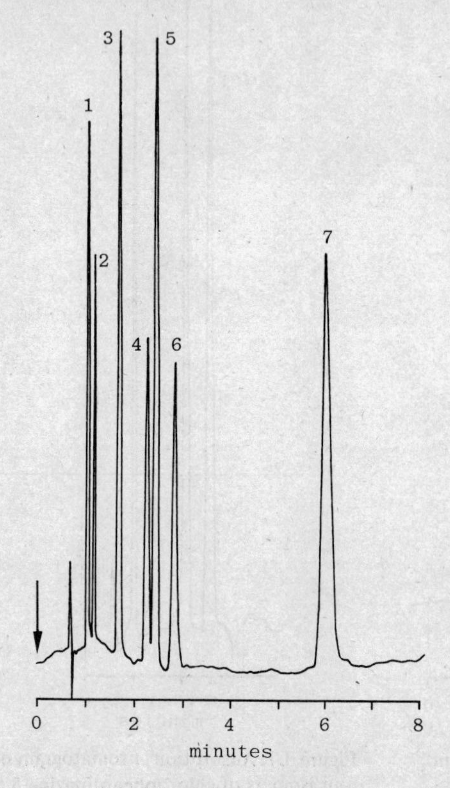

Figure 3.8. Adsorption chromatogram of plant protectants (26).

Column: 250 mm
Packing: 7 μm LiChrosorb SI 60
Eluent: heptane
Pressure: 72 bar
Solutes: (1) 1,3-dichlorobenzene, (2) chlorobenzene, (3) p,p′,-DDE, (4) 4,4,′-DDM, (5) o,p′-DDT, (6) p,p′-DDT, (7) p,p′-DDD
Detector: UV photometer, 254 nm

REFERENCES

1. Tswett, M., *Trav. Soc. Nat. Warsowie 14* (1903) 6.
2. Day, D.T., *Science 17* (1903) 1007.
3. Snyder, L.R., *Principles of Adsorption Chromatography.* Dekker, New York, 1963.
4. Iler, R.K., *Colloid Chemistry of Silica and Silicate.* Cornell University Press, New York, 1955.
5. Snyder, L.R. and Kirkland, J.J., *Modern Liquid Chromatography.* Wiley Interscience, New York, 1974.

6. Saunders, D.L., *J. Chromatogr. 125* (1976) 165.
7. Engelhardt, H., *J. Chromatogr. Sci. 15* (1977) 380.
8. Scott, R.P.W. and Kucera, P., *J. Chromatogr. Sci. 13* (1975) 837.
9. Scott, R.P.W. and Kucera, P., *Anal. Chem. 45* (1973) 749.
10. Majors, R.E., *Anal. Chem. 45* (1973) 755.
11. Done, J.N., *J. Chromatogr. 125* (1976) 43.
12. Scott, R.P.W. and Kucera, P., *J. Chromatogr. 119* (1976) 467.
13. Godbille, E. and Devaux, P., *J. Chromatogr. Sci. 12* (1974) 564.
14. Done, J.N., Kennedy, G.J. and Knox, J.H., in *Gas Chromatography 1972* (ed. S.G. Perry) London, Applied Science Publishers, p. 145. 1973.
15. Messer, W.H. and Unger, R.K., *J. Chromatogr. 149* (1978) 1.
16. Williams, R.C., Baker, D.R., Larmann, J.P. and Hudson, D.R., *Int. Lab.* (Nov/Dec 1973) 39.
17. Kirkpatrick, D., *J. Chromatogr. 121* (1976) 153.
18. Evans, N., Jackson, A.H., Matlin, S.A. and Towill, R., *J. Chromatogr. 125* (1976) 345.
19. Szalonti, G., *J. Chromatogr. 126* (1976) 9.
20. Cadogan, J.I.G., Done, J.N., Lunn, G. and (in part) Lim, P.K.K., *J. Chem. Soc.*, Perkin Transactions I (1976), 1749.
21. Ward, R.S. and Pelter, A., *J. Chromatogr. Sci. 12* (1974) 510.
22. Hulhaven, R. and Desager, J.P., *J. Chromatogr. 125* (1976) 369.
23. Done, J.N. and Knox, J.H., *Spherisorb Applications.* Phase Separations Ltd.
24. Pryde, A., *J. Chromatogr. 152* (1978) 123.
25. Kirkland, J.J., in *Gas Chromatography 1972* (ed. S.G. Perry) p.39. Applied Science Publishers, London, 1973.
26. Merck Applications data sheet 76–1, Darmstadt, W. Germany.

4. LIQUID-LIQUID PARTITION CHROMATOGRAPHY

Liquid-liquid partition chromatography depends upon the different distribution coefficients of the components of a mixture between a stationary and a mobile liquid phase. It was invented in 1941 by Martin and Synge, who use it to separate amino acid derivatives, and it was described in one of the classic papers in the chromatographic literature (1). Their remarkable understanding of chromatography is shown by the following quotation, in which they put forward two principles that were to be the key to improvement of the whole technique of chromatography: "Thus the smallest HETP should be obtainable by using very small particles and a high pressure difference across the length of the column". These suggestions were finally put into practice only with the development of HPLC in the late 1960s, and it is fitting that liquid-liquid partition chromatography is now feeling the benefit of the ideas formulated by the inventors of the technique over 25 years ago.

4.1 THEORY

The theoretical background to liquid-liquid partition chromatography is not as well established or documented as that of adsorption chromatography. It is possible, however, to derive equations with broad similarities to those of adsorption chromatography. Snyder *et al.* (2) have derived the following equation, which describes the distribution coefficient, D, of a solute, i, between a stationary and mobile phase. The parameters involved are the molar volume, V_i, of the solute, i, and the solubility parameters of the solute, stationary phase and mobile phase, δ_i, δ_s and δ_m:

$$\log D = \frac{V_i}{RT}[(\delta_i - \delta_m)^2 - (\delta_i - \delta_s)^2] \quad (4.1)$$

Solubility parameters were first introduced by Hildebrand and co-workers (3,4). They are defined by

$$\delta_i = (\Delta H_{vap}/V_i)^{1/2} \quad (4.2)$$

and are by convention listed in units of (calories $cm^{-3})^{1/2}$. Thus for hexane $\Delta H_{vap} = 28400$ J mol^{-1} = 6800 cal mol^{-1}, $V_i = 131.3$ cm^3 mol^{-1} whence $\delta = 7.20$ $cal^{1/2}$ $cm^{-3/2}$. Equation 4.1 is based upon the assumption that the components of the solution show regular-solution behaviour. In "regular solutions" the heat of mixing is non-zero but the entropy of mixing is ideal. Evidently equation 4.1 can only give a rough value for the distribution coefficient, since such an assumption will not hold accurately for liquids that are so incompatible as to form two-phase mixtures (which require $|\delta_s - \delta_m|$ of 7 or more). Nevertheless the approach can be used for semi-quantitative prediction. Values of δ for many common chromatographic mobile phases and some possible stationary phases for partition chromatography are given in the table in the appendix. The distribution coefficient is related to the capacity factor k' by

$$k' = D(V_s/V_m) \quad (4.3)$$

where V_s is the volume of stationary phase and V_m is the volume of the mobile phase (see chapter 2). Combining equations 4.1 and 4.3 gives

$$\ln k' = \ln(V_s/V_m) + (V_i/RT)\,[(\delta_i - \delta_m)^2 - (\delta_i - \delta_s)^2] \quad (4.4)$$

The application of equation 4.4 to practical chromatography is best illustrated by an example. A useful partition system would have $V_s/V_m \approx 0.4$ corresponding to a loading of 30% of stationary phase by weight on a typical silica gel; that is $\ln(V_s/V_m) \approx -1$. Taking a typical value of V_i as 100 cm^3, and ambient temperature as 300 K, gives $V_i/RT = 1/6$ cm^3 cal^{-1}. Equation 4.4 then becomes

$$\ln k' = -1 + 1/6\,[(\delta_i - \delta_m)^2 - (\delta_i - \delta_s)^2] \quad (4.5)$$

and using typical values of $\delta_m = 7.1$ (e.g. pentane as eluent), $\delta_s = 14.7$ (e.g. triethylene glycol as stationary phase) we can see how the equation functions.

A polar solute might have a solubility parameter of, say, $\delta_i = 13$ and consequently $\log_{10} k'$ would be about 2, i.e. $k' \approx 100$ and the solute would be very strongly retained. A non-polar solute might have a solubility parameter of, say, $\delta_i \approx 8.5$ and hence $\log k'$ would be about -3 and the solute would be unretained. Solutes of intermediate polarity, for example phenol ($\delta_i = 11.4$), would have $\log k'$ near zero or $k' \approx 1$. In general, solutes would be usefully retained only when $[(\delta_i - \delta_m)^2 - (\delta_i - \delta_s)^2]$ is between 6 and 20; for the two liquids chosen this requires δ_i between 11.3 and 12.2, i.e. k' values of between 1 and 10. Thus careful choice of δ_m and δ_s can lead to the achievement of separation, but evidently any two-phase mixture will be able to handle only a very limited range of solute polarities or δ_i values. Further adjustment of k' can then be obtained either by changing the phase ratio (V_s/V_m) or by using different phase systems with the same solubility parameters. The latter method exploits the fact that although liquids can have the same overall solubility parameters, these can be made up from contributions from different effects. For example, a proton-accepting stationary phase and a proton-donating stationary phase might have the same solubility parameter but would have slightly different solubilities for weakly basic solutes. This effect can be considered to be the equivalent of the secondary solvent effects in adsorption chromatography and in a similar fashion can be used to provide variations in selectivity. A detailed discussion of the effects of the different contributions to the solubility parameter is given by Keller *et al.* (5).

4.2 PARTITION SYSTEMS

Liquid-liquid partition chromatography is carried out in various ways. Classical systems employing two simple immiscible liquids are still used but some more specialised systems involving specially designed cocktails for one or other phase

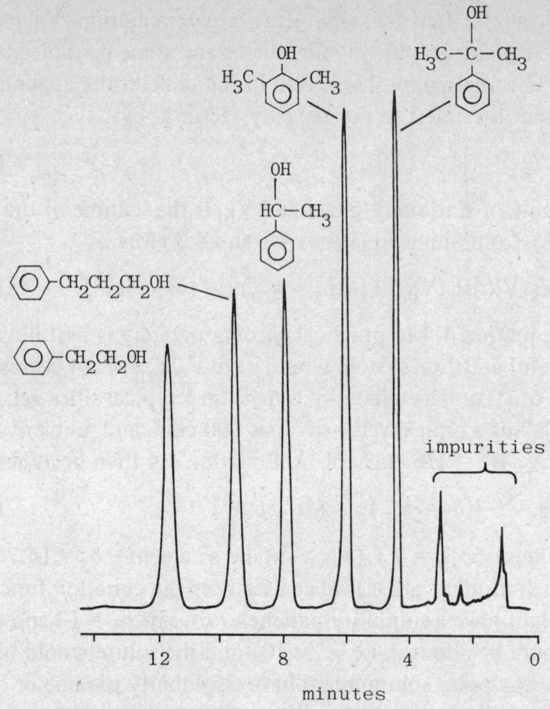

Figure 4.1. Partition chromatogram of aromatic alcohols (7).

Column:	250 x 2 mm
Packing:	6 μm porous microspheres (silica) pore size 35 nm
Eluent:	hexane saturated with β,β′-oxydipropionitrile
Stationary Phase:	β,β′-oxydipropionitrile
Solutes:	as shown
Pressure:	40 bar
Detector:	UV photometer, 254 nm, 0.05 AUFS

have been introduced recently. Some of these will be dealt with below.

Classical partition systems, either normal or reversed-phase, require a column of more or less inert packing material that has been coated with a stationary phase. A typical normal-phase system would use triethylene glycol coated on to a wide pore silica gel as stationary phase and triethylene-glycol-saturated hexane as mobile phase. This type of system has been used to separate a wide range of fairly polar materials such as phenols (6), steroids (6) and aromatic alcohols (7). Figures 4.1 and 4.2 show typical separations. Because of the polar nature of the stationary phase, this type of column is stable as long as the eluent is saturated with the stationary phase. The polar nature of the latter ensures that it is held strongly on to the silica support.

A special type of partition system has been introduced by Huber and co-workers (8–10), which employs a two-phase ternary mixture, formed, for example, by mixing methylene chloride, ethanol and water (8,9), or iso-octane, ethanol and water (10) in appropriate proportions. The three-component

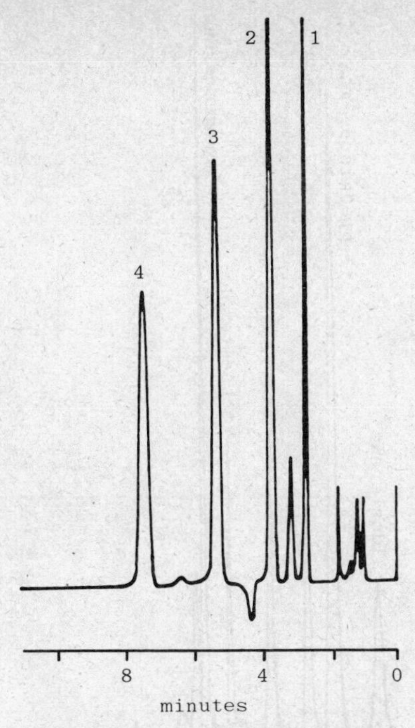

Figure 4.2. Partition chromatogram of phenols on pellicular material (6).

Column:	1000 x 2 mm
Packing:	29 μm Zipax
Eluent:	hexane saturated with β,β′-oxydipropionitrile
Stationary Phase:	2% w/w of β,β′-oxydipropionitrile
Solutes:	(1) unretained solute, (2) 2,4 xylenol, (3) o-cresol, (4) m-cresol, (5) phenol
Detection:	UV photometer, 254 nm, 0.16 AUFS

mixture is stirred at constant temperature until equilibrium is reached. The aqueous layer is separated and used as the stationary phase, while the organic-rich layer is used as mobile phase. Strict temperature control is necessary, but excellent separations have been carried out and good selectivity for similar compounds can also be achieved. An example of the separation of some steroids is shown in figure 4.3.

Reversed-phase systems, in which a non-polar stationary phase (e.g. squalane) is coated on to a silica-based support and the samples eluted with water/alcohol mobile phases, are no longer used. The stability of such systems is low, because the forces holding, say, squalane to even a silylated silica are so weak that the stationary phase is easily washed from the column. A compromise reversed-phase packing material was developed, which had a polymeric hydrocarbon stationary phase on the support, but although quite successful it has now been superseded by the chemically bonded reversed-phase packings dealt with in full in the next chapter.

This brief description has indicated both the strengths and weaknesses of

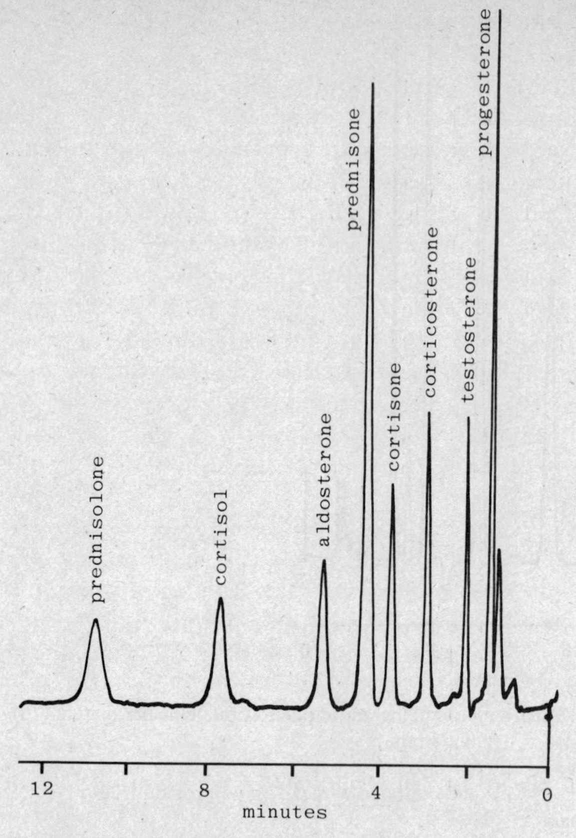

Figure 4.3. Partition chromatogram of steroids using a three-component two-phase system (9).

Column:	300 x 5 mm
Packing:	6 μm Spherosil XOA 400
Eluent:	methylene chloride/ethanol/water (93.6:4.7:1.7 v/v/v) equilibrated with aqueous phase
Stationary Phase:	aqueous phase equilibrated with eluent
Solutes:	as marked
Detector:	UV photometer, 240 nm, 0.25 AUFS

partition chromatography. Since the two phases have to be essentially immiscible, the polarity range of sample that can be separated by any column, as already noted, is relatively small. This is because the sample has to be soluble in both phases and to a certain extent this is an incompatible requirement. However, the wide range of phase systems that can and have been used does allow small differences in solute properties to be exploited to achieve separation. The main disadvantages of partition chromatography are the necessity for saturating the eluent with stationary phase and the closely related disadvantage of not being able to use gradient elution. These are again more reasons for the development of chemically bonded phases for HPLC.

4.3 COLUMN PREPARATION AND USE

As already indicated, the eluent in liquid-liquid partition chromatography must be kept continuously saturated with stationary phase if the latter is not to be stripped from the packing. This is carried out by passing the eluent through a pre-column packed with an inexpensive support coated with between 30 and 40% of the stationary phase. Microparticulate silica columns are usually packed using a slurry method (11; see also chapter 12), and because of this the coating must be applied after the column is packed. Various methods have been described, including equilibration by passing saturated eluent through the column, but the most convenient seems to be one of the *in situ* methods in which the stationary phase (alone or as a 30% solution) is either injected or pumped through the column followed by saturated eluent until equilibration is achieved (12). To determine whether or not equilibrium has been reached the k'-value of a standard compound is determined repeatedly. When k' is constant the column is ready for use. Certain additional measures must be taken if reproducible chromatography is to be obtained. For very accurate work both the pre-column and the analytical column should be thermostatted, and samples should be dissolved in eluent saturated with stationary phase. High flow-rates should be avoided, since mechanical loss or dissolution of the stationary phase can occur under these conditions. In particular the latter can happen because of frictional heating of the eluent as it passes over the column packing.

4.4 APPLICATIONS OF PARTITION CHROMATOGRAPHY

Partition chromatography has been used to separate a wide range of compounds, and some of the examples are collected in table 4.1. These have been chosen to illustrate the types of partition systems that have been used, although some of the separations would now be carried out more conveniently with reversed-phase bonded materials. A survey of the more recent literature has shown that, except for ion-pair partition systems, partition chromatography in the classical mode with a mechanically held stationary phase is little used for routine separations. It would appear that the necessity for the pre-saturation of the mobile phase with stationary phase, and the incompatibility of these systems with gradient elution, is a severe restriction on their use except for special cases. The

Table 4.1. Examples of separations using partition chromatography.

Stationary phase	Eluent	Sample	Ref.
Tris-(2-cyanoethoxy) propane	hexane	2,4-dinitrophenyl hydrazones of aldehydes	13
Polyethylene glycol 400	iso-octane carbon tetrachloride	ethylene oxide oligomers	14
Squalane	water/acetonitrile	hydrocarbons	15
Hydrocarbon polymer	methanol/ phosphoric acid	vitamins	16
β,β'-oxydipropionitrile (BOP)	hexane	phenols	6
Amberlite LA-1	water of pH 11.5	steroids	17
Cyanoethylsilicone polymer	water	coumarins	15

most important of these is normal-phase ion-pair partition chromatography, which is becoming a popular separation technique for ionisable compounds, and which will be dealt with in chapter 6. Nevertheless, the advantages of selectivity inherent in the range of stationary phases available for liquid-liquid partition chromatography should always be borne in mind when selecting the mode of HPLC to use for the separation.

REFERENCES

1. Martin, A.J.P. and Synge, R.L.M., *Biochem. J. 35* (1941) 1358.
2. Karger, B.L., Snyder, L.R. and Horvath, C., *An Introduction to Separation Science.* Wiley Interscience, New York, 1973.
3. Hildebrand, J.H. and Scott, R.L. *Regular Solutions.* Prentice-Hall, Englewood Cliffs, New Jersey, 1962.
4. Hildebrand, J.H. and Scott, R.L. *Solubility of Non-electrolytes,* 3rd. ed. Reinhold Publishing Corp., New York, 1949.
5. Keller, R.A., Karger, B.L. and Snyder, L.R., *Gas Chromatography 1970* (ed. Stock and Perry) p. 125. Institute of Petroleum, London, 1971.
6. Done, J.N., Knox, J.H. and Loheac, J., *Applications of High Speed Liquid Chromatography.* John Wiley, London, 1974.
7. Kirkland, J.J., *J. Chromatogr. Sci. 10* (1972) 593.
8. Huber, J.F.K., Meijers, D.A.M. and Hulsman, J.A.R.J., *Anal. Chem. 44* (1972) 111.
9. Hesse, G. and Hovermann, W., *Chromatographia 6* (1973) 345.
10. Huber, J.F.K., Alderlieste, E.T., Harren, H. and Poppe, H., *Anal. Chem. 45* (1973) 1337.
11. Majors, R.E., *Anal. Chem. 45* (1973) 755.
12. Kirkland, J.J. and Dilks, C.H., *Anal. Chem. 45* (1973) 1778.
13. Papa, L.J. and Turner, L.P., *J. Chromatogr. Sci. 10* (1972) 747.
14. Huber, J.F.K., Kolder, F.F.M. and Miller, J.M., *Anal. Chem. 44* (1971) 105.
15. Schmit, J.A., in *Modern Practice of Liquid Chromatography* (ed. J.J. Kirkland). Wiley Interscience, New York, 1971.
16. Du Pont Liquid Chromatography Methods, Bulletin 820M10, March 1972.
17. Siggia, S. and Dishman, R.A., *Anal. Chem. 42* (1970) 1223.

5. CHEMICALLY BONDED STATIONARY PHASES FOR HPLC

In chapters 3 and 4 the high chromatographic efficiencies obtainable using adsorption or liquid-liquid partition systems were exemplified. However, adsorption chromatography is not usually applicable to the chromatography of very polar or ionic molecules, and liquid-liquid partition chromatography can be troublesome in practice due to the need for a thermostatted pre-column loaded with stationary phase to prevent stripping of the stationary phase from the analytical column. For these reasons, chromatography on chemically bonded stationary phases, prepared by bonding an organic moiety to the surface of an adsorbent, is becoming increasingly popular. The advantages of bonded phases are that polar and ionic molecules can be efficiently chromatographed, that there is much more freedom in the choice of eluent than in liquid-liquid partition chromatography, and that gradient elution techniques can be used without stripping the stationary phase. Their main disadvantage is that the bonded stationary phase can be cleaved off by buffer solutions that are too basic or too acidic (see below) and by oxidising agents.

Virtually all current bonded materials are based upon silica gels. The main types of organic group used in commercial bonded phases are:

(i) Hydrophobic groups, especially octadecyl ($C_{18}H_{37}$) groups but also groups with shorter chain lengths such as C_1, C_2 and C_8.

(ii) Polar groups such as aminopropyl, cyanopropyl, ether and glycol.

(iii) Ion-exchange groups such as sulphonic acid, amino and quaternary ammonium.

Materials for exclusion chromatography have also been chemically reacted with silane reagents to remove unwanted adsorptive effects (1). The preparation and applications of chemically bonded stationary phases in HPLC have been discussed in several review articles (2–8).

5.1 PREPARATION OF CHEMICALLY BONDED STATIONARY PHASES

As indicated in figure 2.8, the surface of silica is covered with a layer of silanol (Si-OH) groups, and reactions at these sites are used to introduce organic groups onto the adsorbent surface. While there are about five OH groups per nm^2 of surface on a silica gel, corresponding to 8–9 $\mu mol\ m^{-2}$, it is stereochemically impossible to react them completely even with groups as small as trimethylsilyl. The maximum surface concentrations of various groupings have been measured by a number of workers and some of their results have been reviewed by Guiochon *et al.* (8). The maximum concentrations of trimethylsilyl, octadecylsilyl and triphenylmethylsilyl are found to be about 4.5, 3.5 and 1.5 $\mu mol\ m^{-2}$ respectively, indicating that at best about half the available OH groups can be reacted. In general the object of derivatization is therefore not to obtain complete substitution of all OH groups but to react a sufficient proportion so that those remaining are inaccessible to solute molecules and do not affect their retention.

The main methods for preparing bonded phases are:

(a) *Reactions with Alcohols and Amines*

The first chemically bonded stationary phases for HPLC were prepared by Halasz and Sebestian (9) by reacting the surface silanol groups with an alcohol, as shown in figure 5.1. Other materials were prepared by converting silica to silica chloride by treatment with thionyl chloride, and then reacting the surface chloride groups with amines (10), as shown in figure 5.2. As these new phases had strands of organic chains pointing away from the silica surface they have become known as "Halasz brushes", and are commercially available from Waters as Durapak supports. A serious disadvantage is their limited hydrolytic stability within the pH range 4–7 only, due to hydrolysis of the Si-OR and Si-NR bonds.

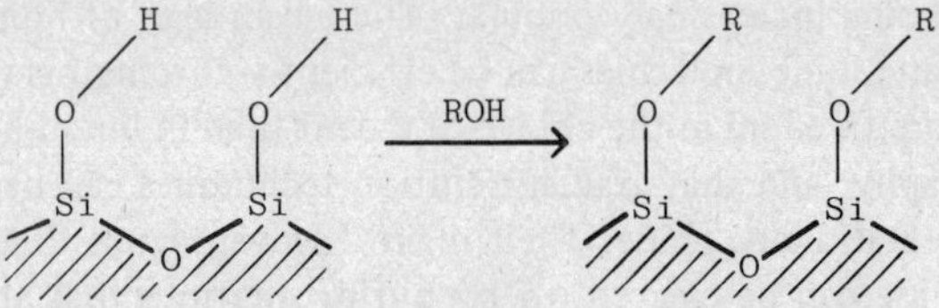

Figure 5.1. Esterification of silanol groups to give an alkoxy silane phase.

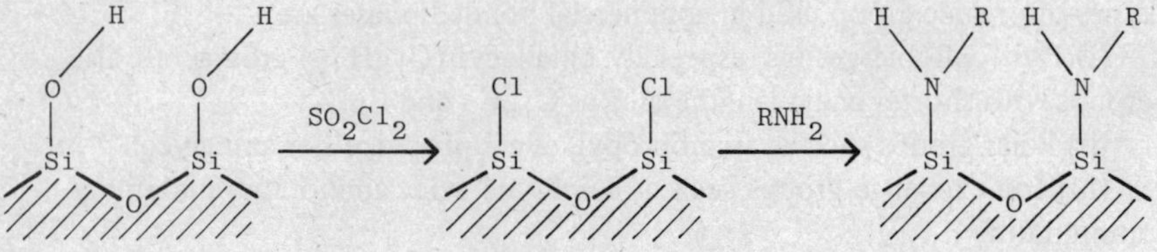

Figure 5.2. Amination of silanol groups to give an N-bonded alkylamino silane phase.

(b) *Reaction with Organosilanes*

The great majority of modern commercially available bonded materials are derived from reactions between organochlorosilanes or alkoxysilanes with the surface silanol groups. Such reactions can be carried out under a range of conditions, as discussed by Majors (11). There are two general approaches to the bonding of organosilanes to silica:

(i) *Bonding under anhydrous conditions.* Here the reaction is carried out under conditions that as far as possible exclude water from the reaction mixture. Typical reaction conditions might involve heating dry silica under reflux with octadecyltrichlorosilane in toluene (11, 12). In the absence of moisture, no hydrolysis of the Si-Cl bonds in the chlorosilane takes place and therefore no polymerisation of the silane. Bonding occurs by elimination of HCl between the organosilane and one or more of the surface silanol groups. After removal of any excess of the silane, the product is hydrolysed to convert unreacted Si-Cl groups to silanol groups, as depicted in figure 5.3.

The presence of these residual silanol groups, arising either from hydrolysis of Si-Cl bonds as shown, or by non-reaction of a surface silanol site, is a complication that may confer chromatographically undesirable adsorptive properties on the material. These residual groups should therefore be "capped", for example with trimethylsilyl groups. This is not always done in practice, and some commercial materials have substantial adsorptive properties due to Si-OH groups. Some of the more widely used commercially available phases based on this reaction are listed in table 5.1.

Clearly a wide range of bonded phases could be prepared by this scheme. The more common materials have octadecyl, cyanoalkyl and phenyl groups. Further reaction, for example sulphonation of phenylalkyl supports to form ion-exchange materials, is frequently carried out (13).

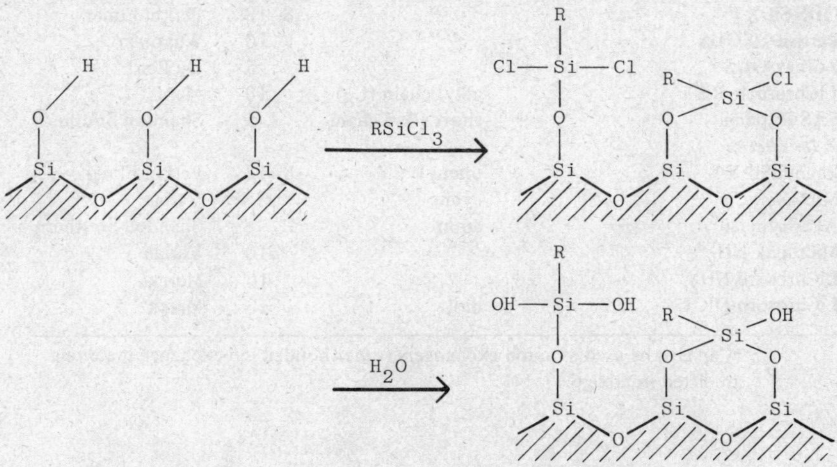

Figure 5.3. Reaction of silanol groups to give an alkyl silyl bonded phase using an alkyl trichlorosilane.

The advantage of supports prepared in this way is that the siloxane (Si-O-Si) bonds formed during the reaction are relatively stable to hydrolysis. Although there is a paucity of detailed information on the stability of bonded phases, they are generally held to be stable in the pH range 3–8. At pH below 2 or 3 the organic groups begin to be cleaved from the support and, at about pH 8, silica itself starts to dissolve. The stability of the bonded phase is also governed by the nature of the bonded group, the composition and ionic strength of the eluent, and the column operating temperature.

(ii) *Reaction with hydrolysis*. As shown in figure 5.4, an alternative way of reacting organosilanes with siliceous surfaces is first to hydrolyse the organochloro- or alkoxysilane to the silanetriol, which partially polymerises. The polymer is then bonded to the support surface by multiple attachments, again via stable siloxane linkages (14, 15). This is the method of preparation of the Du Pont

Permaphases, in which the organic group may be an octadecyl, γ-glycidoxypropyl ("ether") or quaternary ammonium group. The siloxane linkages again confer good hydrolytic stability on these supports.

Table 5.1. Commercial microparticulate bonded reversed-phase and polar supports. (The list is intended as a guide and is by no means complete. A comprehensive list of packing materials for HPLC is given in reference 3.)

Name	Bonded Group	Particle size (μm)	Supplier
Reversed-Phase Materials (non-polar)			
μ-Bondapak C_{18}/Porasil	octadecyl	8–12	Waters
LiChrosorb RP-18	"	5,10	Merck
Micropak-CH	"	10	Varian
ODS-Hypersil	"	5	Shandon Southern
ODS-Sil X-I	"	8–18	Perkin Elmer
Partisil-10 ODS	"	10	Whatman
Zorbax-ODS	"	5	Du Pont
LiChrosorb RP-8	alkyl chain (C_8)	10	Merck
SAS-Hypersil	short alkyl chain	5	Shandon Southern
Polar Phases			
Phenyl-Sil-X-I	phenyl	8–18	Perkin Elmer
Micropak-CN	cyano	10	Varian
APS-Hypersil[a]	amino	5	Shandon Southern
Micropak-NH_2[a]	"	10	Varian
LiChrosorb NH_2	"	10	Merck
LiChrosorb DIOL	diol	10	Merck

[a]Can also be used as anion exchangers (other bonded ion-exchange materials are listed in table 6.1).

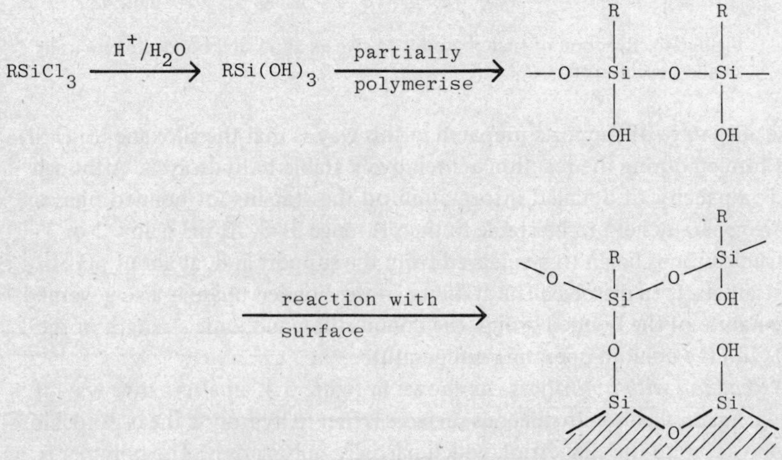

$$RSiCl_3 \xrightarrow{H^+/H_2O} RSi(OH)_3 \xrightarrow[\text{polymerise}]{\text{partially}} \text{—O—Si(R)(OH)—O—Si(R)(OH)—}$$

$$\xrightarrow[\text{surface}]{\text{reaction with}}$$

Figure 5.4. Reaction of silanol groups with a partially polymerised silane to give a polymeric stationary phase.

(c) *Reactions involving Grignard and Organolithium Reagents*
Organic groups can be bonded to the surface of silica via direct Si-C bonds, by reaction of silica chloride with Grignard or organolithium reagents (16–18), as depicted in figure 5.5. These supports also possess good hydrolytic stability.

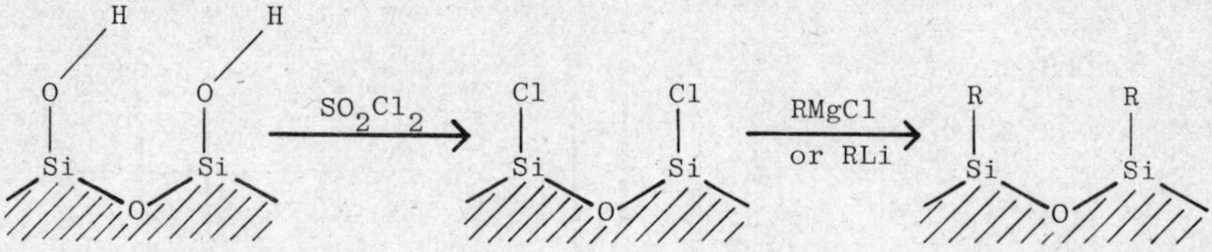

Figure 5.5. Formation of an alkyl silyl bonded phase using a Grignard reagent or an alkyl lithium.

5.2 THE RANGE OF CHEMICALLY BONDED STATIONARY PHASES

As has been demonstrated above, there is virtually no limit to the number of chemically bonded stationary phases that can be prepared. Fortunately only a fairly small number of bonded phases is required to cope with nearly all the types of HPLC separations likely to be encountered in practice. The range would include at most a long-chain hydrocarbon material (e.g. an ODS-silica), possibly a short-chain hydrocarbon or phenyl-bonded material, a polar-bonded material with, for example, a cyano, amino or ether group, a cation exchanger (sulphonic acid) and an anion exchanger (an amine or quaternary ammonium salt). The list is short because selectivity in HPLC can readily be adjusted by variation in the nature of the eluent. This contrasts with the situation in gas chromatography where the eluent has fixed properties and selectivity can be adjusted only by altering the stationary phase, the support, or the temperature. As a simple illustration, an ODS material could be used with a methanol/water eluent containing a long-chain alkyl cyanide. The cyanide would be preferentially partitioned into the hydrophobic ODS surface and the material would thus act as a dynamic cyano phase. In general, the effective nature of the bonded phase can be varied over a wide range by adding to the eluent small quantities of organic amines and acids, electrolytes, long-chain acidic or basic detergents, or complexing agents.

5.3 CHROMATOGRAPHY ON REVERSED-PHASE MATERIALS

Reversed-phase chromatography is complementary to normal-phase adsorption chromatography. The term "reversed phase" implies the use of a polar eluent with a non-polar stationary phase, and it is especially useful for the chromatography of polar molecules. In adsorption chromatography an organic eluent such as hexane or methylene chloride is used along with an adsorbent such as silica or alumina. Molecules of increasing polarity are increasingly retained by the support until, with very polar molecules, a stage is reached where the deactivator (water, methanol, etc.) is displaced, the chromatographic efficiency falls and tailed peaks are obtained. This general problem with very polar molecules can be overcome using reversed-phase chromatography in which a hydrocarbon-bonded support such as ODS-silica is used with an aqueous eluent containing a proportion of, say, methanol or acetonitrile. The more polar solutes now have greater affinities for the eluent and so elute in reversed order of polarity (i.e. most polar first). An example of such a separation is that of paracetamol metabolites shown in figure 5.6.

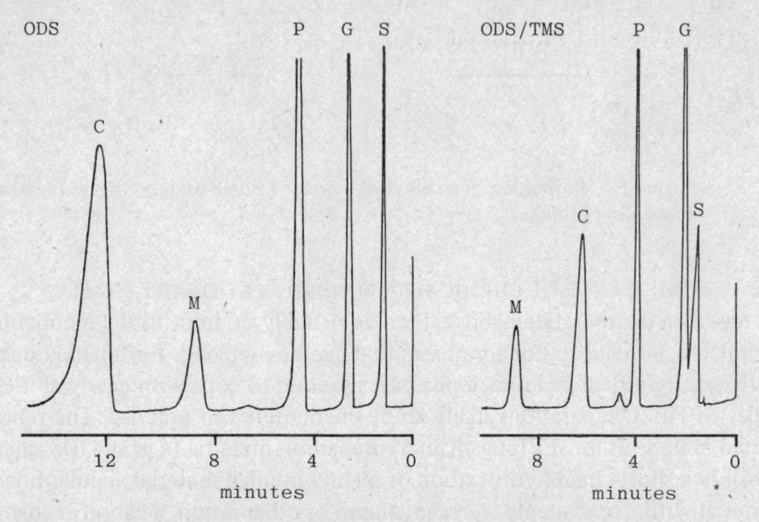

Figure 5.6. Comparative chromatogram of paracetamol metabolites on ODS silica gel and ODS/TMS silica gel, illustrating the improved performance obtained when exposed silanol groups are "capped" (19).

Column:	125 x 5 mm
Packing:	6 μm ODS silica and ODS/TMS silica (Wolfson LC Unit)
Eluent:	water/methanol/formic acid (86:14:0.1 v/v/v)
Solutes:	Paracetamol sulphate (S), paracetamol glucuronide (G), paracetamol (P), paracetamol cysteine (C), paracetamol mercapturic acid (M)
Detector:	UV photometer, 254 nm, 0.1 AUFS

Recent ideas on the mechanism of reversed-phase chromatography are contained in references 6, 20–22. Stated simply, the hydrophobic surface extracts the more lipophilic component of the eluent to form an organic-rich layer at the particle surface in which the chromatographically useful partitioning takes place. The retentions of various members within a class of compounds in reversed-phase chromatography have been correlated to their solubilities in the eluent. Such a correlation has obvious usefulness in the prediction of the structure or polarity of unknown members of a class. Examples of the use of reversed-phase chromatography to study partition coefficients and structure/activity relationships include separations of aromatic hydrocarbons (23), urea herbicides (6), chlorinated pesticides (24), antibiotics (25) and other drugs (26). The mechanism of the separation process on reversed-phase materials is further discussed in the ion-pair partition section (chapter 6).

Although reversed-phase chromatography is suited to the separation of polar molecules, non-polar molecules can also be successfully separated by using an eluent that is sufficiently rich in the organic component (e.g. methanol) to cause the solutes to elute in a reasonable time. This approach may be useful for separating compounds that are poorly resolved by adsorption chromatography or where the sample contains polar impurities that would degrade the performance of an adsorption column. An example of the separation of a mixture of polycyclic aromatic hydrocarbons (PAH) by reversed-phase chromatography is

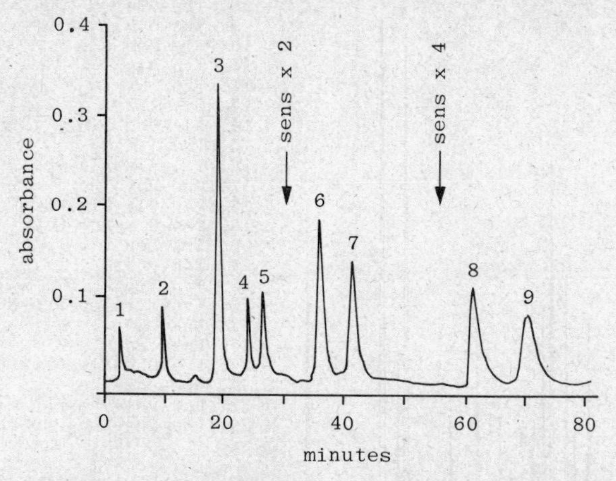

Figure 5.7. Chromatogram of a synthetic mixture of polynuclear aromatic hydrocarbons (PAHs) (27).

Column:	250 x 4 mm
Packing:	Zorbax ODS
Eluent:	methanol/water (65:35 v/v)
Temperature:	60°C
Pressure:	80 bar
Solutes:	(1) solvent peak, (2) naphthalene, (3) anthracene, (4) fluoranthene, (5) pyrene, (6) triphenylene, (7) benzo[a]anthracene, (8) perylene, (9) benzo[a]pyrene
Detector:	UV photometer, 254 nm

shown in figure 5.7 (27). The separation was used in the analysis of PAH in suspended particulate-matter samples.

Generally, chromatography on alkyl-bonded supports (reversed-phase chromatography) is used in conjunction with a mobile phase containing a significant proportion of an organic constituent. However, Horvath and co-workers (28) have recently used ODS-silica materials with aqueous buffer solutions containing no organic component for the separation of ionic or ionisable solutes. Solute retention was influenced by the pH of the eluent, which governed the degree of ionisation of the solute and hence its partition coefficient between the organic stationary phase and the aqueous mobile phase. Varying the ionic strength of the eluent also influenced the solute retention but, whereas in ion-exchange chromatography an increase in ionic strength causes a decrease in k' (see chapter 6), in this system capacity ratios increased with increasing ionic strength. This novel type of chromatography was designated hydrophobic chromatography, and the separation of some aromatic amino acids and small peptides using this technique is shown in figure 5.8.

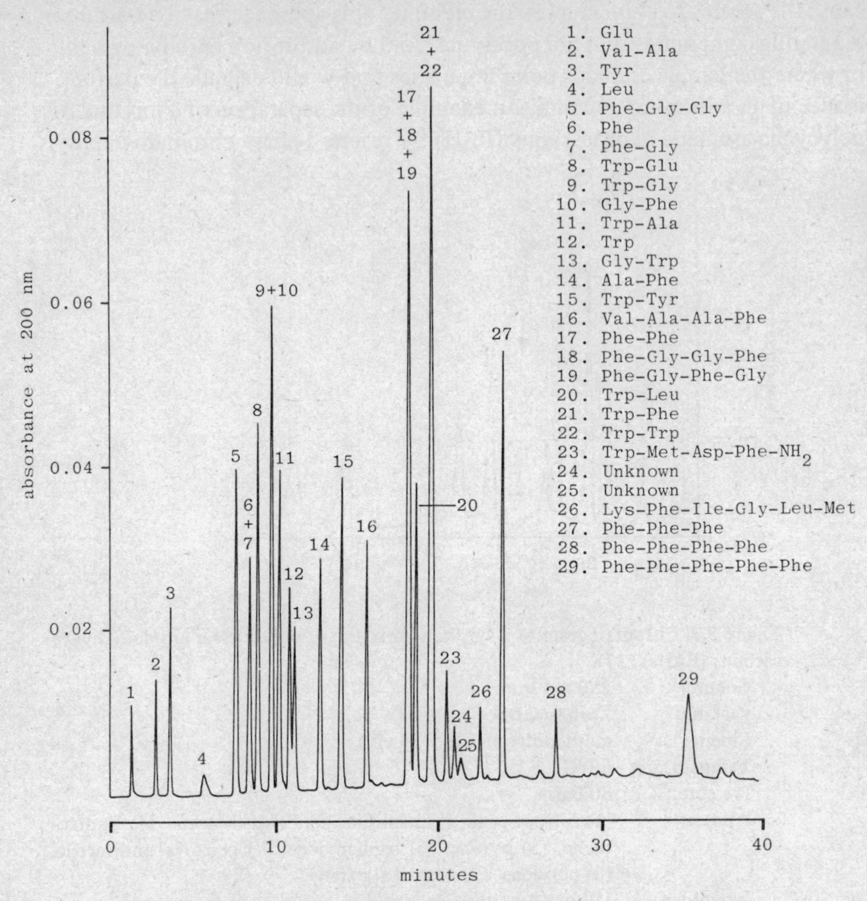

Figure 5.8. Gradient elution chromatogram of amino acids and peptides by hydrophobic chromatography (28).

Column:	250 x 4.6 mm
Packing:	5 μm LiChrosorb RP 18
Eluent:	0.5 M $HClO_4$ with acetonitrile as gradient former
Temperature:	70°C
Pressure:	150 bar
Solutes:	as indicated
Detector:	UV photometer, 200 nm, 0.1 AUFS

5.4 RESIDUAL SILANOL GROUPS IN REVERSED-PHASE SUPPORTS

Since reversed-phase materials are used specifically for the chromatography of polar molecules, any silanol groups that remain accessible to solutes after the bonding are likely to make an important contribution to the chromatography of such solutes. Furthermore, highly polar molecules, which tail badly on the unbonded silica, will interact with these residual silanol groups with deleterious effects. Efforts should therefore be made to ensure that the surface of a reversed-phase material is uniformly hydrophobic, for example, by blocking residual silanol groups with trimethylsilyl groups.

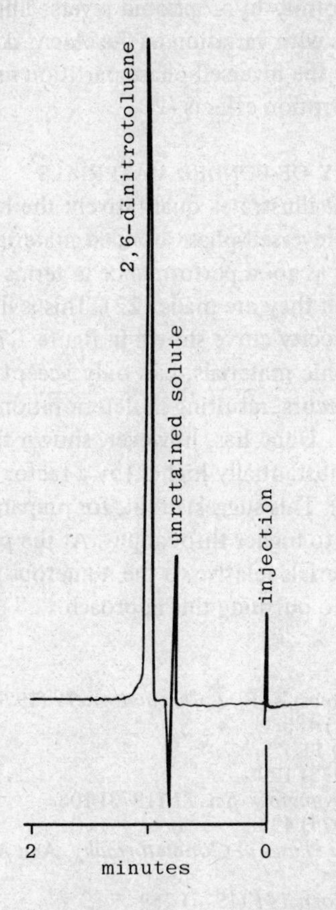

Figure 5.9. Chromatogram of 2,6-dinitrotoluene on SAS Hypersil using dry hexane as eluent, showing absence of significant adsorptive capacity. k′ for solute = 0.25.

The presence of residual silanol groups can be detected most readily by using methyl red indicator (21), which turns red in the presence of acidic silanol groups, but a more sensitive test is to chromatograph a polar solute on the reversed-phase material with dry hexane as eluent. The degree of retention is then a sensitive guide to the presence of residual silanols; if the solute is essentially unretained, the absence of silanols may be assumed. Figure 5.9 shows as an example the chromatography of 2,6-dinitrotoluene on SAS-Hypersil (prepared by bonding short alkyl chains to the surface of silica) using dry hexane as eluent, the support having first been washed with sodium-dried ether and freshly dried hexane. Under these conditions the polar solute had k′ = 0.25. When chromatographed on an adsorbent, this solute would normally require a much more polar eluent to elute it in a reasonable time. Thus the virtual absence of accessible silanol groups may be assumed. In a similar experiment, using another commercially available reversed-phase material, Done showed that about 50% of the original silanol sites were still available after bonding (29).

An incidental advantage of using a uniformly hydrophobic reversed-phase material is that trends in k′ values of solutes with variation in the eluent composition can be more easily explained, since the reversed-phase partition mechanism is not complicated by secondary adsorption effects (20).

5.5 PLATE EFFICIENCY AND CAPACITY OF BONDED MATERIALS

The test chromatogram shown in figure 1.2 illustrates qualitatively the high plate efficiency that can be achieved with reversed-phase bonded materials. Contrary to general belief they have at least as good performance in terms of plate height as do the adsorbents from which they are made (22). This is illustrated quantitatively by the plate-height/velocity curve shown in figure 2.7.

Bonded materials, like all chromatographic materials, can only accept a certain load of sample before overloading occurs, resulting in deterioration of peak shape, peak symmetry and resolution. Done has, however, shown that the capacity of reversed-phase supports is substantially higher (by a factor of up to 10) than that of a range of silica gels (29). This suggests that, for preparative HPLC, reversed-phase materials should lead to higher throughput. At the present time, of course, the expense of bonded materials relative to the numerous inexpensive silica gels is a severe disincentive to pursuing this approach.

REFERENCES

1. Unger, K., Kern, R., Minou, M.C. and Krebs, K.-F., *J. Chromatogr. 99* (1974) 435.
2. Pryde, A., *J. Chromatogr. Sci. 12* (1974) 486.
3. Majors, R.E., *Intl. Lab.* (Nov/Dec 1975) 11.
4. Locke, D.C., *J. Chromatogr. Sci. 11* (1973) 120.
5. Leitch, R.E. and De Stefano, J.J., *J. Chromatogr. Sci. 11* (1973) 105.
6. Locke, D.C., *J. Chromatogr. Sci. 12* (1974) 433.
7. Grushka, E., (editor) *Bonded Stationary Phases in Chromatography*. Ann Arbor Science Publ. Mich., 1974.
8. Colin, H. and Guiochon, G., *J. Chromatogr. 141* (1977) 289.
9. Halasz, I. and Sebestian, I., *Angew. Chem. Int. Ed. 8* (1969) 453.
10. Brust, O-E., Sebestian, I. and Halasz, I., *J. Chromatogr. 83* (1973) 15.
11. Majors, R.E. and Hopper, M.J., *J. Chromatogr. Sci. 12* (1974) 767.
12. Kirkland, J.J., *Chromatographia 8* (1975) 661.
13. Cox, G.B., Loscombe, C.R., Slucutt, M.J., Sugden, K. and Upfield, J.A., *J. Chromatogr. 117* (1976) 269.
14. Kirkland, J.J. and Yates, P.C., US. Patent 3,772,181 (March 1973).
15. Kirkland, J.J., *J. Chromatogr. Sci. 9* (1971) 206.
16. Locke, D.C., Schmermund, J.T., and Banner, B., *Anal. Chem. 44* (1972) 90.
17. Sebestian, I. and Halasz, I., *Chromatographia 7* (1974) 371.
18. Saunders, D.H., Barford, R.A., Magidman, P., Olszewski, L.T. and Rothbart, H.L., *Anal. Chem. 46* (1974) 834.
19. Knox, J..H. and Jurand, J., *J. Chromatogr. 142* (1977) 651.
20. Kikta, Jr. E.J. and Grushka, E., *Anal. Chem. 48* (1976) 1098.
21. Karch, K., Sebestian, I. and Halasz, I., *J. Chromatogr. 122* (1976) 3.
22. Knox, J.H. and Pryde, A., *J. Chromatogr. 112* (1975) 171.
23. Sleight, R.B., *J. Chromatogr. 83* (1973) 31.
24. Seiber, J.N., *J. Chromatogr. 94* (1974) 151.
25. Mechlinski, W. and Schaffner, C.P., *J. Chromatogr. 99* (1974) 619.
26. Twitchett, P.J. and Moffat, A.C., *J. Chromatogr. 111* (1975) 149.

27. Dong, M., Locke, D.C. and Ferrand, E., *Anal. Chem. 48* (1976) 368.
28. Horvath, C., Melander, W. and Molnar, I., *J. Chromatogr. 142* (1977) 623.
29. Done, J.N., *J. Chromatogr. 125* (1976) 43.

6. ION-EXCHANGE AND ION-PAIR CHROMATOGRAPHY

The chromatography of ionisable compounds has long presented problems, particularly in adsorption chromatography where the highly polar nature of the acid and basic functional groups and of the corresponding ions can cause displacement of the protective layer of deactivator that is necessary to ensure homogeneity of the adsorbing sites (see chapter 3 for more detailed discussion). The result is that peaks are severely tailed and resolution poor. The situation is more favourable when using reversed-phase bonded materials, which are, as it were, permanently deactivated with regard to adsorption. The eluent must, however, be buffered in order to control the position of the acid-base equilibria

$$R.COOH + H_2O \rightleftharpoons R.COO^- + H_3O^+ \tag{6.1}$$

$$R.NH_2 + H_2O \rightleftharpoons R.NH_3^+ + OH^- \tag{6.2}$$

If the eluent is not buffered, the proportions of the neutral and ionised forms will change throughout the chromatographic band. Since the two forms of the solute will certainly have different degrees of retention, peak tailing is inevitable. Horvath and co-workers (1), as discussed in chapter 5, have exploited this form of chromatography, which they term "hydrophobic chromatography", and it is now a powerful technique for handling ionisable materials. In their method the neutral form of the solute is more strongly retained than the ionised form, and therefore decreasing the pH for acids, or increasing it for bases, increases the retention. Increasing the ionic strength of the eluent also tends to increase retention.

The more widely used techniques of ion-exchange and ion-pair chromatography work on the opposite principle, namely preferential sorption into an organic phase of the ionised forms of solutes. To achieve this unlikely result they make use of the formation of ion pairs in the organic phase. The ion pairs are formed between the solute ion and an oppositely charged ion, which may be at a fixed site as in classical ion-exchange systems, or may be added into the aqueous phase in the form of a salt as in ion-pair chromatography. Ion-exchange systems were the first to be developed (2) but they are now fairly rapidly being displaced by the more versatile and efficient ion-pair systems (3).

6.1 ION-EXCHANGE CHROMATOGRAPHY

Ion-exchange chromatography (IEC) has a long history of applications, including the analysis of amino acids (2,4), nucleic acid components (5), carbohydrates (6) and, more recently, nucleotides (7).

Typical ion-exchange materials contain either acidic groups such as sulphonic acid or carboxylic acid for the separation of cations, or basic groups such as amine or quaternary amine for the separation of anions. The ion-exchange process may be represented by equation 6.3, which would apply, for example, to an anion exchange material:

$$\sim N^{+},NO_3^{-} + X^{-} \rightleftharpoons \sim N^{+},X^{-} + NO_3^{-} \qquad (6.3)$$

$\sim N^{+},NO_3^{-}$ represents an ion exchange site, the $\sim N^{+}$ group being fixed to the matrix while the associated counter ion, NO_3^{-}, is in the liquid phase; the counter ion can be displaced by a solute ion, X^{-}, to give the ion pair $\sim N^{+},X^{-}$ In classical ion-exchange resins the acidic or basic groups are bonded to a styrene-divinyl benzene copolymer. The classical resins have the disadvantages when applied in HPLC that their volumes change in different solvents due to swelling, they are compressible under high pressures and their mass transfer characteristics are poor. The last defect arises from the fairly high degree of crosslinking (normally 8 to 12%) required to make the resins sufficiently rigid to withstand pressure. The reticulation of such resins is then around 1.5 nm, which is very much less than that of a silica gel as used in adsorption chromatography, which is around 10 nm. Only the smallest molecules can readily penetrate such resins, and even then mass transfer is relatively slow because of the highly structured form of the aqueous phase within the pores of the resin. A substantial improvement in IEC was made in 1960 by Hamilton (4), who was the first to recognise clearly the benefit of using small particles and high pressure. An advance on the original homogeneous resin beads was made by Horvath, Preiss and Lipsky in 1967 (8) with the introduction of pellicular ion-exchange materials, in which glass beads were coated with a thin layer of ion-exchange resin. They were followed by the Zipax ion exchangers (9), in which polymer was deposited in a porous layer on a glass bead. These materials were incompressible and their mass transfer properties were somewhat better than those of homogeneous resin beads of the same size, since the solute molecules had a shorter distance to travel within the resin. The pellicular materials are now, however, mainly of historical interest, and the highest efficiencies are obtained with homogeneous bonded materials in which the ion-exchange groups are chemically bonded to microparticulate fully porous silica gels (10,11). Examples of separations on such materials are shown in figure 6.1, and may be compared to a similar separation on a classical ion-exchange material of about the same particle size shown in figure 6.2.

In general the efficiencies obtained with the new bonded ion exchangers are very similar to those with adsorbents and reversed-phase materials of the same particle size. In the past ion-exchange separations were often carried out at elevated temperatures in order to speed mass transfer and reduce the effects of the high viscosities of aqueous buffer solutions, but the bonded materials give excellent efficiencies at ambient temperature. Examples of commercially available ion exchangers are listed in table 6.1.

As shown above, the ion-exchange process involves the displacement of a counter ion, which must also be present in the eluent, by a solute ion. For effective chromatography, as indicated in chapter 2, this displacement reaction

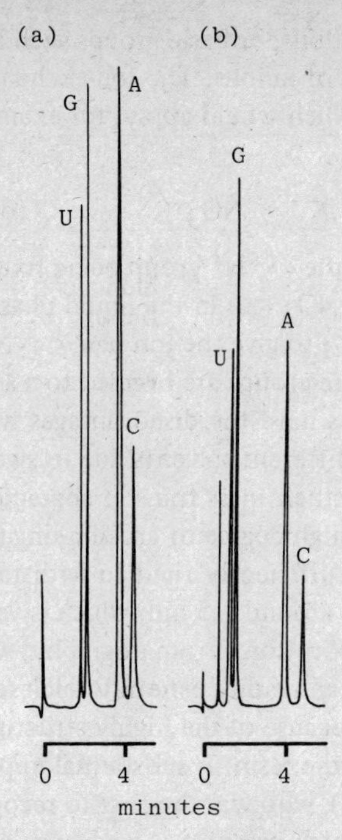

Figure 6.1. Chromatogram of nucleosides on bonded cation exchangers (10).

Column:	150 x 4.6 mm
Packing:	(a) phenyl ethyl sulphonic acid bonded to 5 μm Merckosorb SI 60 (b) same bonded to 5 μm Spherisorb S5W
Eluent:	0.05 M ammonium formate in water/ethanol (90:10 v/v), pH 4.8
Temperature:	50°C
Solutes:	uridine (U), guanine (G), adenosine (A), cytidine (C)
Detector:	UV photometer, 254 nm

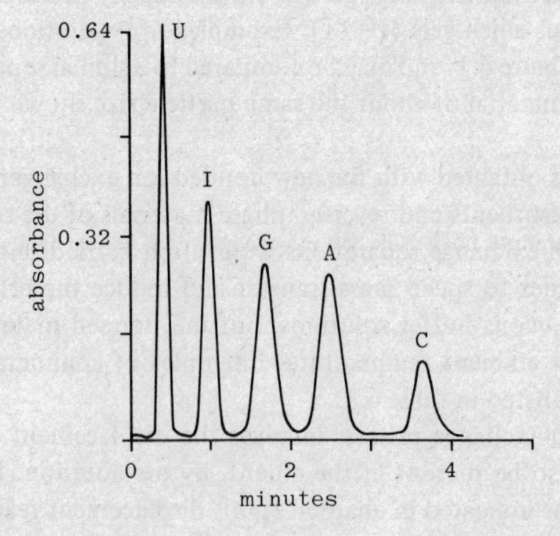

Table 6.1. Commercially available ion-exchange materials. (The list is representative only: a more comprehensive list is given in ref. 3 of chapter 5.)

Name	Description	Particle size (μm)	Supplier
Anion Exchanger[a]			
Aminex A-28	porous polymer beads	7–11	Bio-Rad
DA-X8A	"	6–10	Durrum
Cation Exchanger[a]			
AA-15	"	16–28	Beckman
Aminex A-7	"	7–11	Bio-Rad
AN-90	"	16–28	Hamilton
DC-4A	"	6–10	Durrum
Anion Exchanger			
AS-Pellionex-SAX	polymer-coated glass beads (pellicular)	44–53	Whatman
Pellicular Anion	"	40	Varian
Zipax SAX	"	25–37	Du Pont
Cation Exchanger			
HS-Pellionex-SCX	"	44–53	Whatman
Pellicular Cation	"	40	Varian
Zipax SCX	bonded pellicular	25–37	Du Pont
Anion Exchanger			
APS-Hypersil[b]	porous silica gel, siloxane bonded	5	Shandon Southern
Micropak-NH_2[b]	"	10	Varian
Partisil-10 SAX	"	10	Whatman
LiChrosorb AN	"	10	Merck
Zipax AAX	bonded pellicular	25–37	Du Pont
Perisorb AN	polymer-coated glass beads (pellicular)	30–40	Merck
Cation Exchanger			
Partisil-10 SCX	porous silica gel, siloxane bonded	10	Whatman
LiChrosorb KAT	"	10	Merck
Perisorb KAT	bonded pellicular	30–40	Merck

[a] Unless otherwise stated, cation exchangers have a $-SO_3H$ functionality, and anion exchangers have a $-N^+R_3$ functionality.
[b] Bonded group is a primary amine. These materials are also used as polar adsorbents.

Figure 6.2. Chromatogram of nucleosides on microparticulate ion-exchange resin beads (12).

Column: 250 x 2.4 mm
Packing: 3–7 μm cation exchange resin
Eluent: 0.4 M ammonium formate, pH 4.8
Temperature: 55°C
Pressure: 325 bar
Solutes: uridine (U), inosine (I), guanine (G), adenosine (A), cytidine (C)
Detector: UV photometer, 254 nm, 0.64 AUFS

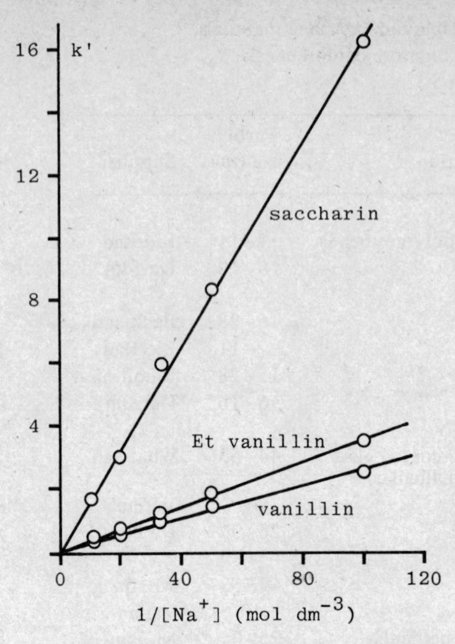

Figure 6.3. Dependence of k′ upon reciprocal of counter ion concentration for elution of acids from a strong anion exchanger, Zipax SAX, by a borate buffer pH 9.2 (13).

must be at equilibrium, and retention is therefore controlled by the equilibrium constant for reaction 6.3, K_{IE}, defined by

$$K_{IE} = \frac{[NO_3^-]\,[\sim N^+,X^-]}{[X^-]\,[\sim N^+,NO_3^-]} \tag{6.4}$$

The capacity k′, being proportional to the distribution coefficient D_{IE} (see chapter 2), is given by

$$k' \propto D_{IE} = \frac{[\sim N^+,X^-]}{[X^-]} = K_{IE}\frac{[\sim N^+,NO_3^-]}{[NO_3^-]} \tag{6.5}$$

Since the concentration of the ion exchange sites $[\sim N^+,NO_3^-]$ is constant and fixed by the structure of the matrix, k′ is inversely proportional to the concentration of the counter ion in the eluent, i.e.

$$k' \propto [NO_3^-]^{-1} \tag{6.6}$$

Figure 6.3 shows an example of such dependence, taken from the work of Nelson (13).

The pH of the eluent also affects k′, although only indirectly through the fraction of solute ionised. From equation 6.1 this is given for an acid by

$$\text{fraction ionised} = \frac{[RCOO^-]}{[RCOO^-] + [RCOOH]} = \frac{K_a}{[H_3O^+] + K_a} \tag{6.7}$$

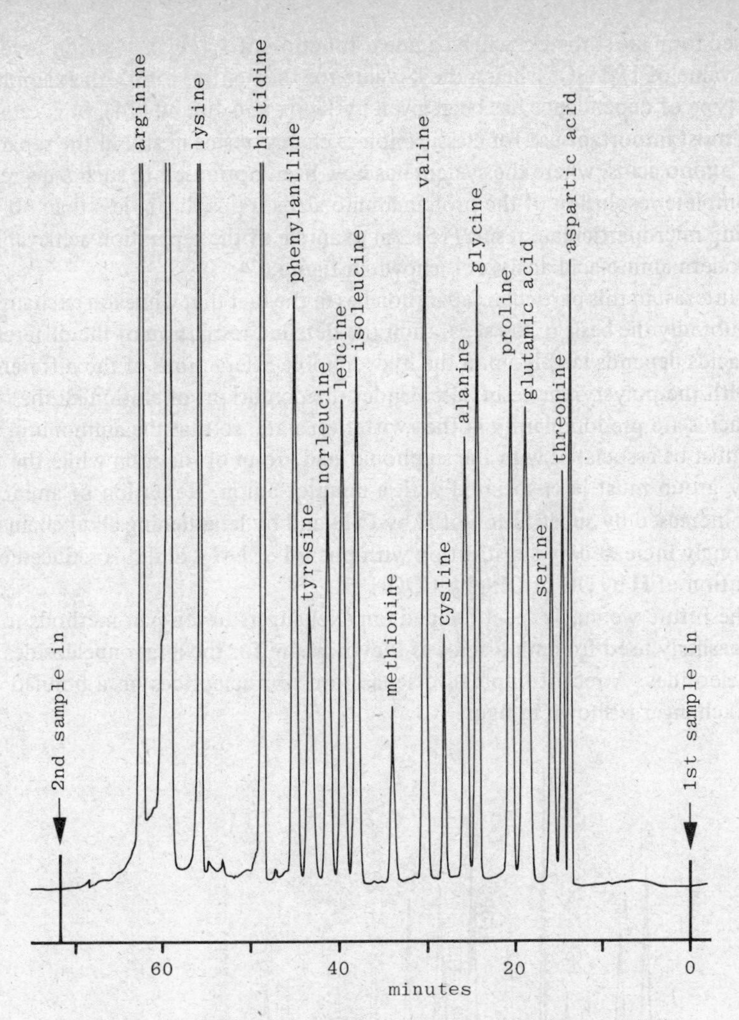

Figure 6.4. Ion-exchange chromatogram of protein amino acids (15).

Column:	350 x 2.6 mm
Packing:	8 μm strong cation exchange resin (spherical)
Eluent:	0.2M sodium citrate/borate buffer with optimised pH gradient from pH 2.2 to pH 11.5
Solutes:	as indicated
Pressure:	40 bar
Temperature:	60°C
Detection:	Post-column reactor followed by fluorimetric monitor. The eluted amino acids are reacted at room temperature with o-phthalaldehyde in presence of 2-mercaptoethanol in a coil reactor. The fluorescent derivative is determined by its emission at 479 nm with excitation at 365 nm

When the pH is significantly less than pK_a the fraction ionised is inversely proportional to $[H_3O^+]$, and so therefore is k', provided that the unionised form is assumed to be unadsorbed by the matrix. When both the ionised and unionised form are sorbed k′ will be a linear function of $1/[H_3O^+]$, the intercept at zero value of $1/[H_3O^+]$ being the k′ value for the neutral form. An example of this type of dependence has been given by Knox and Jurand (14).

The most important use for classical ion-exchange resins in still in the separation of amino acids, where the system has now been optimised to such a degree that complete resolution of the protein amino acids is possible in less than 60 min using microparticulate resins (15). An example of the separation achievable on a modern amino-acid analyser is shown in figure 6.4.

The interest in this particular separation lies in the fact that while ion exchange is undoubtedly the basis of the separation the detailed resolution of the different amino acids depends largely upon the hydrophobic interactions of the different acids with the polystyrene resin base. Under the conditions of elution all the amino acids are predominantly in the zwitterion state, so that the ammonium group must be associated with the sulphonic acid group of the resin while the carboxy group must be associated with a counter anion. Retention of amino acids is increased by substitution of H by CH_3 and by lengthening alkyl chains; it is strongly increased by substitution with phenyl or NH_2; and it is reduced by substitution of H by OH or CH_3 by COOH.

In the future we may expect bonded ion exchangers or ion-pair methods to be increasingly used for amino acids as they now are for the larger nucleosides and nucleotides. A recent impressive separation of nucleotides on a bonded anion exchanger is shown in figure 6.5.

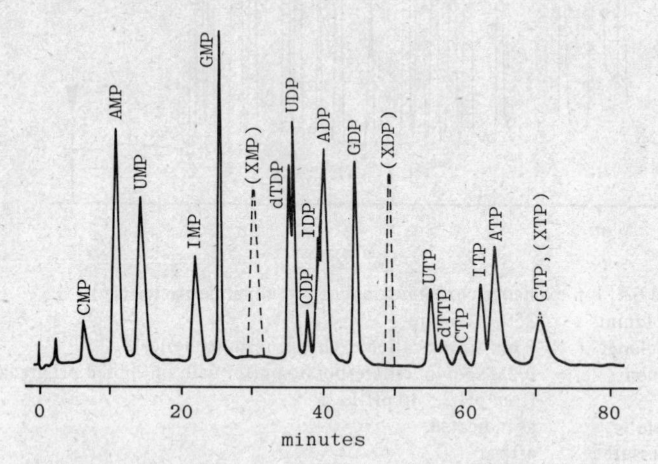

Figure 6.5. Gradient elution chromatogram of mono-, di- and triphosphate nucleotides on a bonded strong anion exchanger (7).

Column: 250 x 4.6 mm
Packing: 10 μm Partisil SAX
Eluent: start 0.007 M KH_2PO_4, pH 4.0
finish 0.25 M KH_2PO_4, 0.50 M KCl, pH 4.5
Solutes: as marked
Detector: UV detector, 254 nm, 0.08 AUFS

6.2 ION-PAIR CHROMATOGRAPHY

Ion-pair extraction has long been exploited for the extraction of drugs from body fluids into organic phases, and the technique has recently been reviewed by Schill (16). In its simplest form the partition equilibrium can be represented by

$$B^{+}_{aq} + P^{-}_{aq} = (B^{+}P^{-})_{org} \tag{6.8}$$

where B^+ may, for example, be the protonated form of the base it is desired to extract, and P^- is the anion of an acid that is to be used as the pairing ion. (B^+P^-), the ion pair, will behave like a very polar organic molecule, and will therefore preferentially dissolve in a polar organic phase (say a higher alcohol/chloromethane mixture) leaving the ionic forms in the aqueous phase. By choosing a suitable pairing ion and adjusting its concentration the base can be very efficiently extracted into the organic phase. Obviously the roles of the anion and cation can be reversed, so that anionic species can be extracted by cationic pairing ions. The large body of data that has been accumulated on such ion-pair equilibria can be exploited with great effect in chromatography, where retention is based upon the establishment of equilibria between a fixed phase or zone and a mobile phase or zone. We can envisage normal-phase ion-pair chromatography in which the aqueous phase is used as the stationary phase (being deposited on a porous support) and the organic phase is the eluent, or reversed-phase ion-pair chromatography where the organic phase forms the stationary phase and the aqueous phase is the eluent. But, in addition, ion pairs can be separated by adsorption chromatography, say, by using silica gel as packing material and eluting the solutes as ion pairs in an organic eluent; or one can use a reversed-phase bonded material to sorb ion pairs from an aqueous eluent where the solute is present in the ionic form. Of these four methods the last is rapidly becoming the most popular, being the simplest to operate although the least simple to interpret in terms of independently measured extraction coefficients.

Although numerous complications can arise in relation to ion-pair extraction, because of the formation and dissociation of ion pairs in the two phases, a simple theory based upon the equilibrium 6.8 is adequate to explain most of the effects observed in liquid chromatography. From equation 6.8 an extraction equilibrium constant E_{BP} may be defined:

$$E_{BP} = \frac{[B^{+}P^{-}]_{org}}{[B^{+}]_{aq}[P^{-}]_{aq}} \tag{6.9}$$

We assume that the pairing ion, P^-, is present at relatively high concentration in the aqueous phase and is associated with a counter ion, C^+, whereas the solute ion, B^+, is always present in very small concentration compared to P^-. The distribution coefficient of B^+ is then given by

$$D_{B+} = \frac{[B^{+}P^{-}]_{org}}{[B^{+}]_{aq}} = E_{BP}[P^{-}]_{aq} \tag{6.10}$$

Since k' is proportional to D_{B^+} in reversed-phase chromatography, or to $1/D_{B^+}$ in normal-phase chromatography, we observe that k' will be proportional to the pairing ion concentration in the former and to its reciprocal in the latter. Control of the pairing ion concentration is thus a simple way of controlling retention.

Qualitatively, D_{B^+} will be increased by making the pairing ion, P^+, larger and more hydrophobic and by improving the solvating power of the organic phase, say by increasing the alcohol content; it will be decreased by adding organic modifiers to the aqueous phase.

If the pairing ion is very hydrophobic, as it will be for detergent ions, P^- will itself be strongly extracted into the organic phase along with its normal counter ion C^+. We then have to consider the additional equilibrium

$$P^-_{aq} + C^+_{aq} = (P^-C^+)_{org} \tag{6.11}$$

Subtracting this equation from 6.8 gives

$$(P^-C^+)_{org} + B^+_{aq} = (P^-B^+)_{org} + C^+_{aq} \tag{6.12}$$

This equation is strikingly similar to 6.3, which applied to fixed site ion-exchange materials, and indeed where the pairing ion is strongly partitioned into the organic phase under normal conditions the ion-pair equilibrium is most simply considered as equivalent to dynamic ion exchange, although there are important differences that should be recognised. The equilibrium constant for reaction 6.12 is

$$K_{IE} = \frac{[P^-B^+]_{org}\,[C^+]_{aq}}{[P^-C^+]_{org}\,[B^+]_{aq}} \tag{6.13}$$

from which the distribution coefficient for B^+ can be written

$$D_B = \frac{[P^-B^+]_{org}}{[B^+]_{aq}} = K_{IE}\,\frac{[P^-C^+]_{org}}{[C^+]_{aq}} \tag{6.14}$$

Whereas in the ion-exchange equilibria with fixed site ion exchangers the concentration $[P^-C^+]_{org}$ was fixed, this is no longer the case in the ion-pair systems. The ratio $[P^-C^+]_{org}/[C^+]_{aq}$ is determined by the equilibrium 6.11. Under conditions where the pairing species is very dilute in both phases this ratio will be constant for any given $[P^-]_{aq}$, and this is indeed the case considered above where $[P^-]_{aq}$ is used directly to control the distribution and retention. At the opposite extreme, where the organic phase is saturated with the ion pair $[P^-C^+]$, the situation is similar to that with fixed site ion exchangers. D_{B^+} is then proportional to $1/[C^+]_{aq}$. This situation will arise in two cases. The first is when the organic phase is essentially pure (P^-C^+), which occurs when a hydrophobic amine like trioctylamine (TAO) is deposited as a stationary phase and an aqueous perchlorate solution used as eluent. The interaction of the eluent and the amine then converts the entire stationary phase into trioctylammonium perchlorate ion pairs (17). The second is when a bonded reversed-phase packing is employed and a very hydrophobic ion is added to the eluent. P^-C^+ ion pairs then form at the interface of the bonded layer and the aqueous phase and a saturated layer is formed (18). Obviously all possibilites between these extremes are possible; in this intermediate regime there is no simple linear relationship between the distribution coefficient and the concentrations of the ions in the aqueous phase.

The technique of ion-pair chromatography in its various forms has great versatility, and this gives it substantial advantages over fixed site ion-exchange systems. It may also be noted that since all the equilibria are dynamic, mass transfer is likely to be rapid in the ion-pair systems. This results in consistently high chromatographic efficiencies, which are in general difficult to attain with fixed site ion exchangers.

In summary, we may distinguish the main practicable modes of ion-pair chromatography:

(a) Normal-phase ion-pair partition chromatography, with an aqueous stationary phase and an organic eluent

(b) Reversed-phase ion-pair partition chromatography, with an organic stationary phase and an aqueous eluent

(c) Reversed-phase ion-pair exchange chromatography, where a strongly hydrophobic pairing ion is employed either as the stationary phase itself or adsorbed onto a bonded stationary phase.

(d) Ion-pair adsorption chromatography.

(a) *Normal-Phase Ion-Pair Partition Chromatography*

This was the first form of ion-pair chromatography to be developed. Eksborg, Lagerstrom, Modin and Schill (19) employed a stationary phase of an aqueous solution of dimethylprotriptylenium to separate various aromatic acids such as benzoic acid, salicylic acid, etc. The eluent was a mixture of hexane/chloroform/pentanol (75:20:5 v/v/v). Later, Persson and Karger (20) used an aqueous perchlorate solution to separate catecholamines, as did Knox and Jurand (21). A rather elegant technique can be exploited in the normal-phase mode, because a UV-absorbing pairing ion can be incorporated in the stationary phase to enable otherwise UV-transparent solutes to be detected as they emerge from the column as UV-absorbing ion pairs. For this purpose picrate or naphthalene sulphonate (22) are convenient ions for use in conjunction with bases, while a tricyclic tranquilliser ion, as used originally by Lagerstrom *et al* (19), is useful for acidic components.

The technique has, unfortunately, some disadvantages. Loading of the column and maintenance of the phase ratio, as with all liquid-liquid partition systems, requires care and accurate thermostatting of the entire system. Two methods of column loading have been used. In the first, stationary phase is pumped through the dry column and the excess held between the particles of packing is displaced by eluent. It emerges as globules in the eluent, and the column is ready for use only when no further bleed of stationary phase is noted. When loading in this way one must take care that the detector does not become contaminated with the stationary phase, which may be difficult to wash out. The alternative method is to inject stationary phase into the column when the eluent is flowing until there is no further change in k′ values or until excess of stationary phase emerges from the column as globules. In all cases the eluent

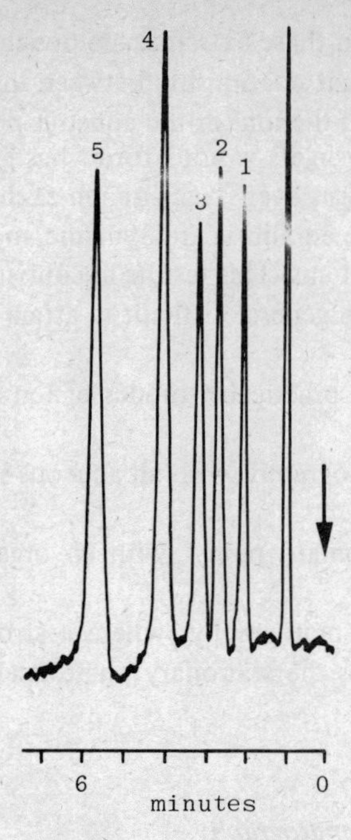

Figure 6.6. Ion-pair partition chromatogram of dipeptides (22).

Column:	150 x 4.5 mm
Packing:	10 μm LiChrospher SI 100
Eluent:	chloroform/pentanol (19:1 v/v)
Stationary Phase:	0.1 M naphthalene-2-sulphonate in water pH 2.3
Solutes:	(1) leu-leu, (2) phe-val, (3) val-phe, (4) leu-val, (5) met-val
Detector:	UV photometer, 254 nm, 0.1 AUFS

must be saturated with the stationary phase.

For reasons that are not immediately apparent the efficiencies obtained in two-phase ion-pair systems are generally lower than those obtained with adsorbents or bonded stationary phases. An illustration of what can be achieved by this method is shown in figure 6.6.

(b) *Reversed-Phase Ion-Pair Partition Chromatography*
Pentanol deposited on Chromosorb was used by Wahlund (23) as the stationary phase for separation of acids, with tetrabutylammonium as the pairing ion added to the aqueous eluent. More recently Schill and co-workers have used pentanol held by a bonded octadecyl silica (24) for the separation of acids. An example of the type of separation obtained is shown in figure 6.7.

(c) *Reversed-Phase Ion-Pair Exchange Chromatography*
Early examples of this form of ion-pair chromatography were the separation of members of the ephedrine class by Schill and co-workers (19) using 0.1 M dioctyl hydrogen phosphate dissolved in chloroform as stationary phase. This ion pair is almost completely insoluble in water. Kraak and Huber (17) used a similar technique in which trioctylamine perchlorate was used as the stationary phase without any diluent to separate sulphonic acids. Not long after this Knox

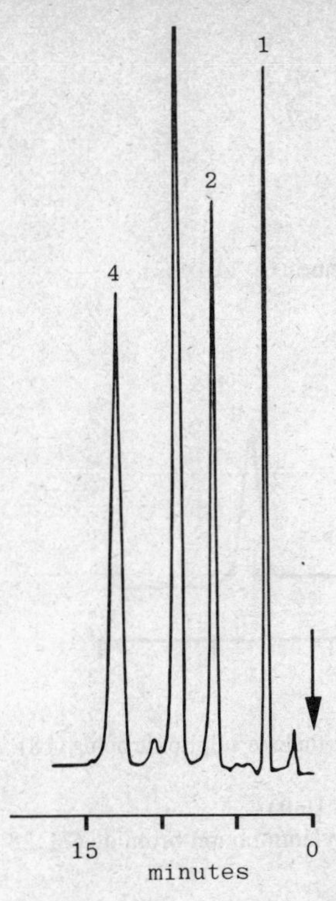

Figure 6.7. Reversed-phase ion-pair partition chromatogram of barbiturates.

Column:	200 x 4.4 mm
Packing:	10 μm LiChrosorb RP2
Eluents:	0.01 M tetrabutylammonium pH 7.7
Stationary Phase:	butyronitrile
Pressure:	27 bar
Solutes:	(1) diethyl-, (2) diallyl-, (3) allylisopropyl-, (4) phenylethyl-barbituric acids
Detector:	UV photometer, 254 nm, 0.02 AUFS

and Laird (18) developed the technique known as "soap chromatography", whereby a reversed-phase bonded silica is used to extract a detergent from an aqueous eluent. They employed cetyltrimethylammonium as the pairing ion, and showed that this was extracted to give what amounted to a monolayer on the surface of the bonded support, a result confirmed independently by Kissinger (25). Using a mixture of propanol and water as eluent Knox and Laird were able to separate previously unresolved sulphonic acids. An example of this type of separation is shown in figure 6.8.

The general technique of using bonded phases is now becoming popular, especially for the separation of drugs and their metabolites. Knox and Jurand (26) developed a highly efficient separation of catecholamines using a variety of anionic detergents as pairing species, and have more recently applied the method to the determination of paracetamol metabolites in urine (27). A chromatogram of a urine sample taken after a therapeutic dose of paracetamol

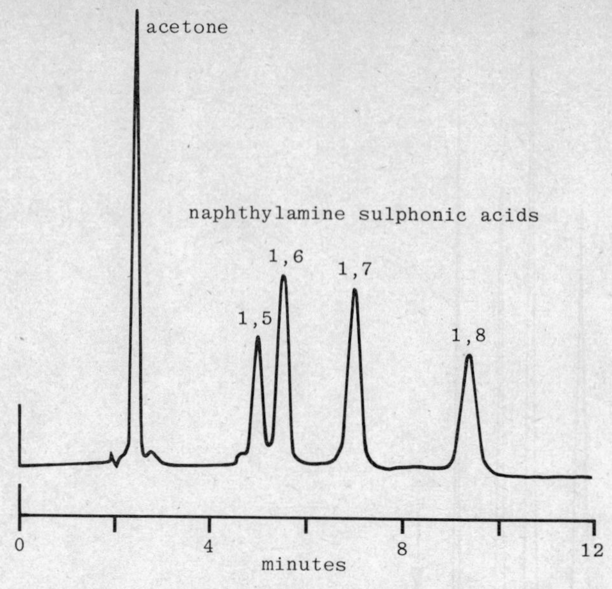

Figure 6.8. Ion-pair chromatogram of naphthalene sulphonic acids (18).
Column: 125 x 5 mm
Packing: 7 μm SAS silica (Wolfson LC Unit)
Eluent: water/propanol/cetyltrimethylammonium bromide (71:28:1 v/v/w)
Solutes: as marked
Detector: UV photometer, 244 nm

is shown in figure 6.9, which demonstrates the excellent peak sharpness that can be obtained by the reversed-phase technique using a bonded support. The plate efficiencies obtainable are in no way inferior to those obtainable in the simpler chromatographic modes.

(d) *Ion-Pair Adsorption Chromatography*

The formation of an ion pair is a convenient way of reducing the polarity of an ionic substance, and thereby bringing it into the range of polarity that can be handled by adsorption chromatography. Although there are relatively few examples of the use of this method it was indeed the first method used by Knox and Laird (18) for the chromatography of the highly ionised sulphonic acids. Figure 6.10 shows the separation of several acids by this method; no discrete liquid stationary phase was present and the acids could not be retained in their ionised forms without detergent addition. The formation of ion pairs for the purpose of "protection" of the ionic groups is in some respects analogous to the adjustment of pH to ensure that acids or bases are in their unionised forms. The difference might be stated as being that the H^+ ion is an extremely strong ion-pairing agent whereas ions such as cetyltrimethylammonium and tetrabutylammonium are rather weaker.

In this same vein one can envisage the use of complexing agents as a means of modifying the sorptive and partitioning behaviour of ions. This method has indeed been used by Karger (28), and its exploitation will undoubtedly open up further novel separation methods.

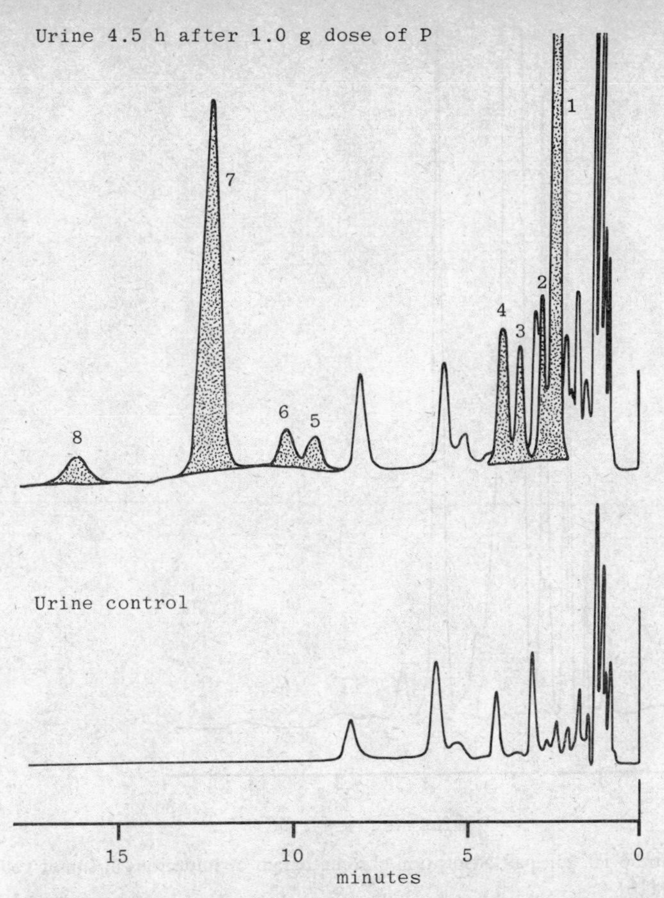

Figure 6.9. Ion-pair chromatogram of urine sample showing paracetamol metabolites taken 4.5 hours after a therapeutic dose (upper) compared with control urine sample (lower) (27).

Column:	125 x 5 mm
Packing:	6 μm ODS/TMS silica (Wolfson LC unit)
Eluent:	water/methanol/formic acid (86:14:0.1 v/v) containing 0.7 mg dm^{-3} of dioctylamine and 2.5 g dm^{-3} KNO_3
Metabolites from 2 mm^3 injection:	(1) paracetamol glucuronide, (2) paracetamol cysteine, (3) paracetamol, (4) methoxy paracetamol, (5) unknown, (6) paracetamol mercapturic acid, (7) paracetamol sulphate, (8) unknown.
Detector:	UV photometer, 249 nm, 0.1 AUFS

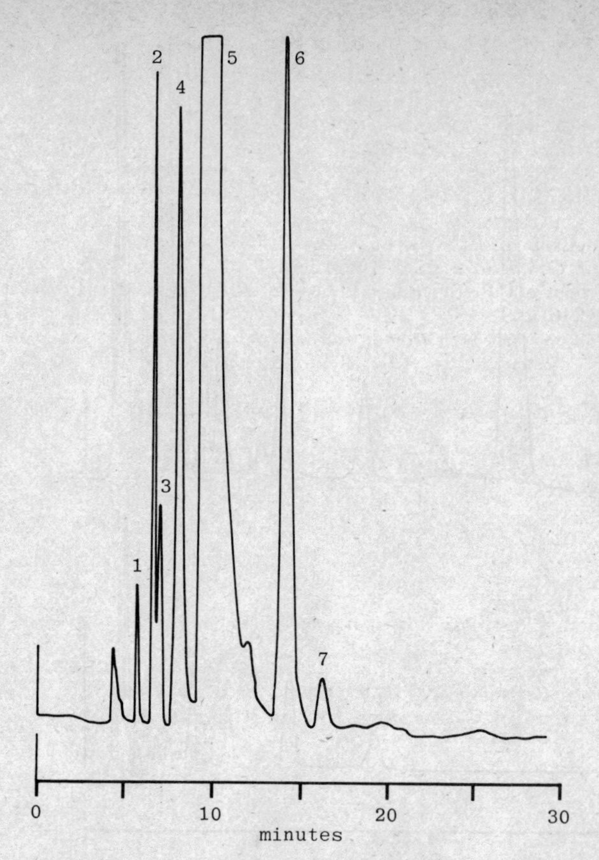

Figure 6.10. Ion-pair adsorption chromatogram of impurities in sunset yellow dye (18).

Column: 125 x 5 mm
Packing: 6.5 μm Partisil
Eluent: propanol/methylene chloride/water/cetyltrimethylammonium bromide (56:32:10:2 v/v/v/v/w)
Solutes: (1) Schaeffer's acid, (2) unknown, (3) sulphanilic acid, (4) triazine (internal standard), (5) sunset yellow, (6), (7) unknown
Detector: UV photometer, 254 nm

REFERENCES

1. Horvath, C., Melander, W. and Molnar, I., *Anal. Chem. 49* (1977), 142.
2. Moore, S. and Stein, W.H., *J. Biol. Chem. 192* (1951) 663.
3. Eksborg, S. and Schill, G., *Anal. Chem. 45* (1973) 2092.
4. Hamilton, P.B., *Anal. Chem. 32* (1960) 1779.
5. Cohn, W.E., *Science 109* (1949) 377.
6. Khym, J.X. and Zill, L.P., *J. Amer. Chem. Soc. 73* (1951) 2399.
7. Hartwick, R.A. and Brown, P.R., *J. Chromatogr. 112* (1975) 651.
8. Horvath, C., Preiss, B.A. and Lipsky, S.R., *Anal. Chem. 39* (1967) 1422.
9. Kirkland, J.J., *J. Chromatogr. Sci. 7* (1969) 361.
10. Cox, G.B., Loscombe, C.R., Slucutt, M.J., Sugden, K. and Upfield, J.A., *J. Chromatogr. 117* (1976) 269.
11. Unger, K.K. and Nyamah, D., *Chromatographia 7* (1974) 63.
12. Gere, D.R., in *Modern Practice of Liquid Chromatography* (ed. Kirkland) p.417. John Wiley, New York and London, 1971.
13. Nelson, J.J., *J. Chromatogr. Sci. 11* (1973) 28.
14. Knox, J.H. and Jurand, *J. Chromatogr. 87* (1973) 95.
15. Rank Hilger, Margate, Kent. Private communication.
16. Schill, G., *Ion Exchange and Solvent Extraction*, Vol. 6 (ed. Marinsky, J.A. and Marcus, Y.) pp. 1–57. Marcel Dekker, New York, 1974.
17. Kraak, J.C. and Huber, J.F.K., *J. Chromatogr. 102* (1974) 333.
18. Knox, J.H. and Laird, G.R., *J. Chromatogr. 122* (1976) 17.
19. Eksborg, S., Lagerstrom, P-O., Modin, R. and Schill, G., *J. Chromatogr. 83* (1973) 99.
20. Persson, B-A. and Karger, B.L., *J. Chromatogr. Sci. 12* (1974) 521.
21. Knox, J.H. and Jurand, J., *J. Chromatogr. 103* (1975) 311.
22. Crommen, J., Fransson, B. and Schill, G., *J. Chromatogr. 142* (1977) 283.
23. Wahlund, K-G., *J. Chromatogr. 115* (1975) 411.
24. Fransson, B., Wahlund, K-G., Johansson, I.M. and Schill, G., *J. Chromatogr. 125* (1976) 327.
25. Kissinger, P.T., *Anal. Chem. 49* (1977) 883.
26. Knox, J.H. and Jurand, J., *J. Chromatogr. 125* (1976) 89.
27. Knox, J.H. and Jurand, J., *J. Chromatogr. 149* (1978) 297.
28. Cooke, N.H.C., Viavattene, R.L., Ekstein, R., Wong, W.S., Davies, G. and Karger, B.L., *J. Chromatogr. 149* (1978) 391.

7. EXCLUSION CHROMATOGRAPHY

Exclusion chromatography or gel-permeation chromatography as generally understood is a technique for separating large molecules on the basis of their partial exclusion from a porous matrix, the reticulation of the matrix being of a similar size to that of the molecules being separated (1). At its simplest, molecules too large to enter the pores elute with the mobile zone in a volume V_o (see chapter 2) while very small molecules, which can completely permeate the matrix, elute in a volume V_m, the volume of the eluent phase. Molecules of intermediate size comparable to the size of the pores in the matrix elute in volumes between V_o and V_m (2). Figure 7.1 shows a high-performance exclusion chromatogram obtained using as support a silica gel with a pore size of around 10 nm.

Figure 7.1 also illustrates the continuity between exclusion and retentive chromatography: benzene elutes at a volume V_m, the high-molecular-weight polystyrene PS 2,700,000 is excluded, PS 20,800 and PS 4000 are partially excluded, while m-dinitrobenzene and acetophenone are retained. Since exclusion on the basis of size can be modified by adsorption, any theory of exclusion based entirely upon molecular and pore geometry can be only an approximation. Fortunately, the experimental conditions are generally arranged to ensure that adsorption is negligible.

7.1 TERMINOLOGY FOR EXCLUSION CHROMATOGRAPHY

In a packed column for exclusion chromatography we can denote the interstitial volume outside the particles as V_o (see chapter 2) and a pore volume within the particles as V_p. The volume of the eluent phase within the column is then

$$V_m = V_o + V_p \tag{7.1}$$

The degree of permeation of a solute is denoted by K, the fraction of the pore volume accessible to the solute in question. The retention volume of any solute is thus expressed by

$$V_R = V_o + KV_p \tag{7.2}$$

Evidently completely excluded solutes have $K = 0$, while totally permeating solutes have $K = 1$. Equation (7.2) may be compared with the corresponding equation for retentive chromatography given previously in chapter 2:

$$V_R = V_m (1 + k') \tag{7.3}$$

or, working with zone rather than phase volumes,

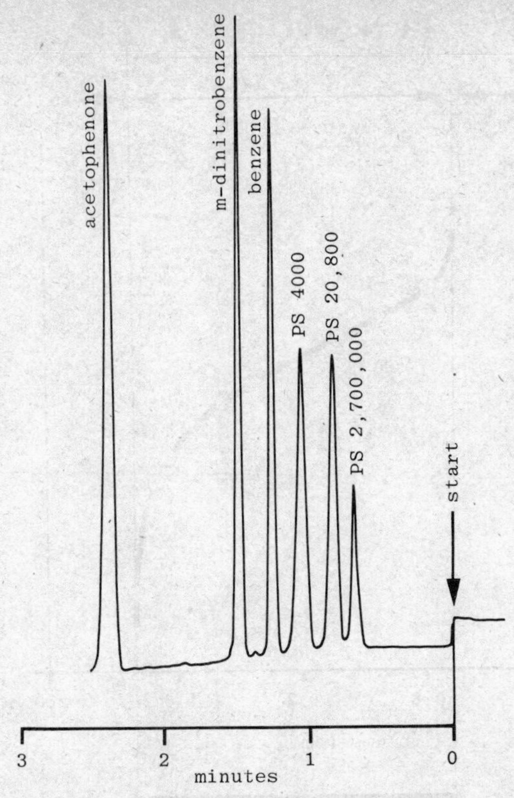

Figure 7.1. High-performance chromatogram of excluded and retained solutes (F. McLennan).

Column: 101 x 7 mm
Packing: 6 μm Hypersil (spherical silica gel of 200 m^2 g^{-1} area)
Eluent: methylene chloride
Solutes: as marked, where PS 4000, for example, is a polystyrene standard with a molecular weight of 4000. Benzene is the totally permeating unretained solute, the PS standards are excluded, while m-dinitrobenzene and acetophenone are retained
Detector: UV photometer, 254 nm, 0.1 AUFS

$$V_R = V_o (1 + k'') \qquad (7.4)$$

from which it is seen that

$$k' = (K-1)V_p/V_m \qquad (7.5)$$

$$k'' = K(V_p/V_m) \qquad (7.6)$$

k' is negative for partially excluded solutes, while k'' falls in the range zero to (V_p/V_m) and is always positive.

The normal way of plotting exclusion data is in the form of a semi-logarithmic calibration curve relating molecular weight to either V_R or K, and a typical such curve for a single matrix material is shown in figure 7.2. The figure also

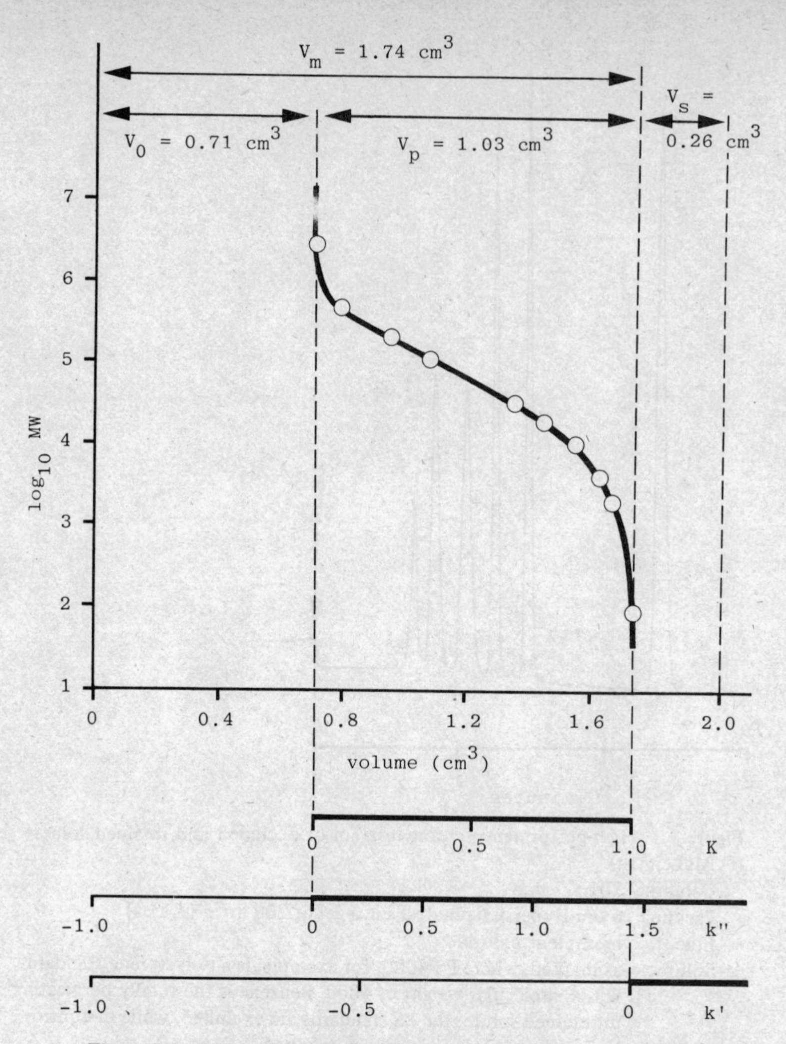

Figure 7.2. Calibration curve for a microparticulate spherical silica gel showing the connection between elution volume V_R and the various distribution ratios K, k″ and k′.

Column: 115 x 4.7 mm (volume 2.00 cm^3)
Packing: SG60 spherical porous silica (experimental batch supplied by Materials Preparation Unit, AERE, Harwell). Mean pore size 60 nm, surface area 100 $m^2\ g^{-1}$, particle porosity 80%
Eluent: methylene chloride
Solutes: polystyrene standards
Detector: UV photometer, 254 nm

indicates the equivalent values of K, k″ and k′, and the volume of the solid part of the matrix, V_s.

Such calibration curves normally have a fairly sharp exclusion limit at the high molecular weight end, a more or less linear intermediate portion, and a gradual curve away from the linear towards the region of total permeation. The

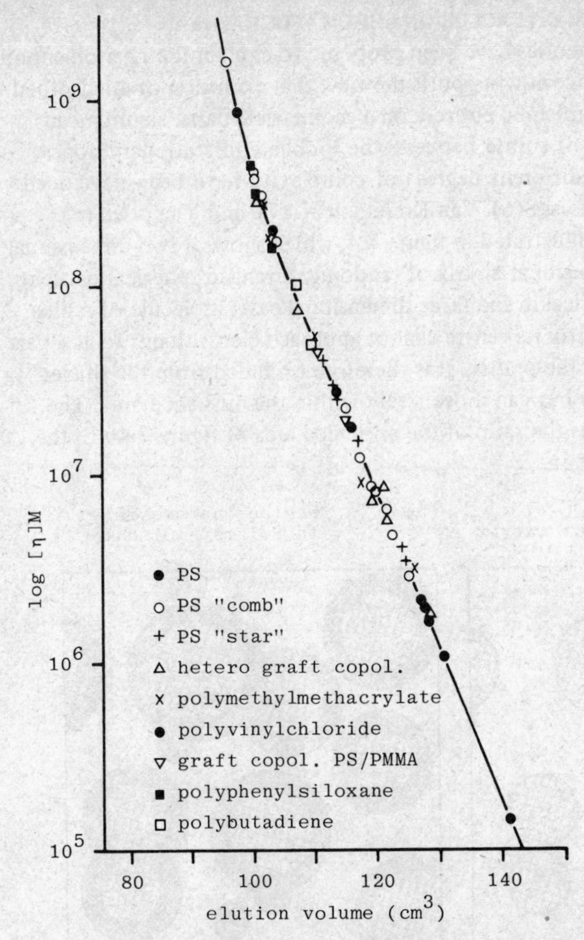

Figure 7.3. Universal calibration curve for a set of four Styragel columns (Waters Associates) using a range of polymers of different types and structures. Note the large elution volume typical of classical GPC systems. Eluent, tetrahydrofuran. (After Grubsic *et al.* (4).)

extent of the linear region generally covers about 1.5 orders of magnitude of M. In order to cover wider ranges of M, combinations of columns are used, each column covering a different molecular weight range (3).

Different types of molecules give different calibration curves because of differences in their molecular configuration but to a good approximation these differences can be accommodated by plotting log $(M[\eta])$ against V_R or K, where $[\eta]$ is the intrinsic viscosity of the polymer. Figure 7.3 shows a universal calibration curve of this type obtained by Grubisic *et al.* (4). A set of four Styragel columns was used in this work to cover 4 orders of magnitude of M.

7.2 THE MECHANISM OF EXCLUSION CHROMATOGRAPHY

Although a number of theories have been proposed to explain the phenomenon of exclusion all the evidence now supports the view that exclusion of unadsorbed large molecules can be explained entirely on a geometrical basis, assuming an equilibrium distribution of solute between the mobile and stationary zones. Geometrical theories of different degrees of complexity have been developed by Giddings *et al.* (5), Casassa (6), Van Krefeld *et al.* (7), and Yau *et al.* (8). The basis of their ideas is illustrated in figure 7.4, which shows a two-dimensional analogue of a three-dimensional matrix of randomly arranged spherical particles. When a circular (or spherical in the three-dimensional case) molecule of radius r is placed within the matrix its centre cannot approach closer than r to any part of the internal surface of the matrix. It is therefore excluded from the shaded area shown in figure 7.4b but can move freely within the unshaded area. The permeation ratio K is thus the ratio of the unshaded area of figure 7.4b to the unshaded area of figure 7.4a.

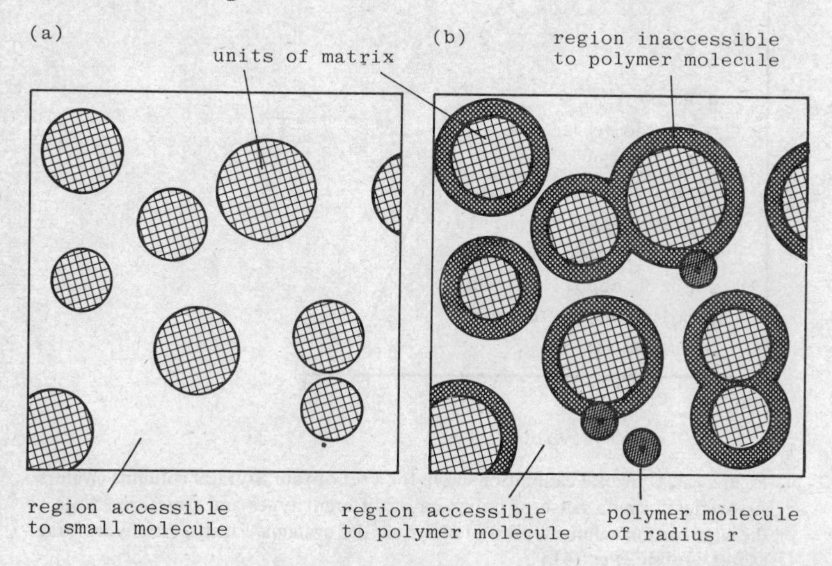

Figure 7.4. Principle of exclusion on a two-dimensional representation: (a) structure of matrix showing solid circular units (shaded) and free space (unshaded); (b) matrix with polymer molecules of radius, r, showing region from which the polymer molecules are excluded. For the structure shown, K $\approx$ 0.6.

Casassa (5) has shown that there is virtually no difference in the calibration curves that are predicted using the model of simple spherical molecules, and those predicted by a much more complex model that assumes polymer molecules to behave as random linear chains undergoing Brownian motion, if the radius of the spherical molecule is taken as about 0.75 times the true hydrodynamic radius of the random chain. Van Krefeld and Van den Hoed (7) used a similar assumption to develop an exact quantitative correlation between experimental data, obtained for polystyrene samples eluted from Porasil, and theoretical calculations. For their model they assumed that the matrix was composed of randomly arranged overlapping spheres of equal radius, while the polymer molecules were taken to be spheres with radii 0.89 times their mean hydraulic radius. This radius r (in Ångstroms) was related to the molecular weight of the polymer by the equation $r = 0.123\ M^{0.588}$. By measuring the pore size and pore volume of the experimental material, and adjusting the model to give the same pore volume as the experimental material, they were able, with no adjustable parameters, to predict the experimental calibration curve within the limits of accuracy of the experimental data.

In view of the weight of evidence there now seems no doubt that exclusion on the basis of size, where adsorption effects are absent, can be explained on an entirely geometrical basis.

7.3 BAND DISPERSION IN EXCLUSION CHROMATOGRAPHY

Until recently very little reliable data on band dispersion in exclusion chromatography had been obtained, and the plate efficiency of columns was generally stated in terms of the plate number for a fully permeating solute such as benzene or styrene. Generally peaks for even the best polymer standards are substantially wider than those for small fully permeating solutes (see figure 7.1) and this has led to the belief that exclusion chromatography is intrinsically less efficient in terms of HETP than other forms of chromatography. There is, however, nothing in the theory of chromatography, as outlined in chapter 2, that would support this view, and indeed Knox and McLennan (9) have shown that the major factor in determining peak widths in most published exclusion chromatograms of polymer standards has almost certainly been the polydispersity of the standard rather than the kinetic broadening of the column.

The polydispersity, P, is defined by

$$P = \frac{M_w}{M_n} \qquad (7.7)$$

where M_w and M_n are the weight- and number-averaged molecular weights. Knox and McLennan showed that even if P was as low as 1.01 a column with a true plate efficiency of 10,000 would give a peak whose width, when used in equation 2.6, would give an apparent value of N of only 3800, whereas if P were 1.1 then the apparent plate efficiency would be only 440. Recent unpublished work by McLennan, in which the effect of polydispersity can be allowed for, indicates that with a truly monodisperse polymer the plate height arising from kinetic spreading processes is indeed the same for partially excluded polymeric solutes as for fully permeating materials. If this is so, it will be possible to use fairly simple unfolding procedures to allow for kinetic spreading when deriving molecular weight distributions from exclusion chromatograms of polymer samples (see reference 10 for a full discussion).

7.4 EQUIPMENT AND PACKING MATERIALS

Size-exclusion chromatography using microparticulate packings is carried out using the same equipment as all other forms of HPLC. Classical columns, which were introduced by Moore (3), were long and relatively wide, typically 6 ft long

Table 7.1. Microparticulate packings for exclusion chromatography.

Name	Type	Particle size (μm)	Solvent compatibility		No. of pore sizes available	Exclusion limits (MW) or pore sizes (Å)	Manufacturer
			aqueous	organic			
AR gel	cross-linked polystyrene	10		yes	9	$10^2 - 10^6$	Applied Chromatography Systems
Bio-beads S	" "	10		yes	6	$4 \times 10^2 - 10^6$	Bio-Rad Laboratories
μ-Bondagel-E	chemically modified silica	10	yes	yes	4	$2 \times 10^3 - 2 \times 10^6$	Waters Associates
CPG	controlled-pore glass	5–10	yes	yes	6	$10^2 - 15 \times 10^5$	Corning
CPG-Glycophase	chemically modified controlled-pore glass	5–10	yes	yes	6	$10^2 - 15 \times 10^5$	Corning
HSG	cross-linked polystyrene	10		yes	8	$4 \times 10^2 - 10^7$	Shimadzu
LiChrospher	silica	10	yes	yes	4	$6 \times 10^4 - 4 \times 10^6$	Merck
SEC	silica	10	yes	yes	3	100, 500 and 1000 Å	Du Pont
Shodex	cross-linked polystyrene	10		yes	6	$10^3 - 5 \times 10^7$	Showa Denko KK
μ-Styragel	" "	10		yes	6	$700 - >10^6$	Waters Associates
Spherosil	silica gel	7	yes	yes	5	80, 140, 360, 680 and 1000 Å	Rhone-Poulenc
TSK gel	cross-linked polystyrene	8–10		yes	7	$50 - 4 \times 10^8$	Toyo Soda Co.

and 3/8 in outside diameter. In order to get adequate resolution three or four such columns were connected in series, and this meant that at least two hours were required to carry out the chromatography. Modern columns are 250 mm long and 5–8 mm internal diameter, but it is still often necessary to have two or three columns in series to achieve the separation. The total elution time is, however, reduced to between 10 and 15 minutes, since in true exclusion chromatography nothing can elute after V_m.

Table 7.1 collects together the details of most of the microparticulate packings currently available for size-exclusion chromatography. They are essentially small versions of materials already used in gel-permeation chromatography. The two predominant types are cross-linked polystyrenes and inorganic packings based on silica gel or glass. The silicas normally used for adsorption chromatography can also be used for size-exclusion systems. Some of these have been studied (11, 12) and separations of very high efficiency (250,000 plates) have been carried out, although total elution times were rather high (11). Adsorption on these silicas has also to be controlled,which may be done by selection of the eluent (cf. figures 2.10 and 7.1), but in more extreme cases protective functional groups must be chemically bonded onto the surface, so that true size-exclusion chromatography is attained. Cross-linked polystyrene cannot normally be used with aqueous mobile phases and because of this current emphasis is on the inorganic packings. The polystyrene packings do, however, have the advantage that smaller molecules can be separated on them, and for this reason they are still important.

The choice of packing depends on the size of the molecules to be separated and the compatibility of the packing for the chosen mobile phase. The columns are packed as for the other forms of HPLC by one of the methods described in chapter 12, then washed through with the mobile phase prior to use for chromatography. Identification of the components is easy if they are discrete compounds, but with polymeric substances calibration must be carried out if a molecular weight distribution is required. Calibration is preferably done with characterised samples of the same polymer but a universal calibration procedure (figure 7.3) can be used if such samples are not available. The detector most used in size-exclusion chromatography is the refractive index monitor. Although it is not often used in other forms of HPLC due to its low sensitivity, this is usually adequate for exclusion chromatgraphy because minimal dilution of the sample occurs. In addition, many polymers do not absorb UV radiation, so a bulk property detector is required. For the separation of discrete molecules of intermediate molecular weight or for UV-absorbing polymers, however, the UV-detectors are becoming more widely used when appropriate.

A final, very important, practical point must be emphasised. The characterisation of polymers by size-exclusion chromatography requires a high degree of control of the mobile phase flow rate (12). The total "working volume" of a series of modern columns might be as little as 10 cm^3. The molecular weight change over this same volume could be by a factor of 10^5, so that each cm^3 of mobile phase corresponds to a three-fold range of molecular weight. This calls for measurement of V_R to within about 1% to obtain a precision of 10% on molecular weight, and precise control and/or measurement of the flow rate through the columns. Computer handling of the data is also required to gain the benefit of the modern packings used in size-exclusion chromatography.

7.5 APPLICATIONS OF SIZE-EXCLUSION CHROMATOGRAPHY

The use of microparticulate packings for the characterising of polymers is now well established. Since most work of this type has previously been done on cross-linked polystyrene packings, this trend has continued with similar microparticulate materials. The operation parameters for this type of system have recently been reviewed (13). The effects of flow rate, sample injection volume and concentration were all evaluated for a set of μ-Styragel columns. The use of silica or glass-based materials for this type of characterisation is likely to become more widespread, and another recent review (14) has shown the potential of this type of material. Seven polystyrene standards were well separated on a set of μ-Bondagel columns in about 24 minutes, and various water-soluble polymers were analysed equally rapidly. Some typical results on a four-column set are shown in figure 7.5, which includes chromatograms from the analysis of two cellulose films on silanised porous silica microspheres using dimethylsulphoxide as mobile phase. The successful use of dimethylsulphoxide, with a viscosity of 2.24×10^{-3} N s m^{-2} (2.24 centipoise), shows that with modern techniques quite viscous mobile phases can be used at room temperature without the necessity of very high operating pressures or temperatures.

The subsequent treatment of the data from these high-speed separations is the same as that for classical systems (10) except that, as already mentioned, complete automation of the system becomes almost essential. A recent review of the possibilities of this type of system described the type of equipment that will be necessary to achieve this (15).

A very important area in which separation on a size basis is widely used is biochemistry. Many separations are slow because of the lack of a rigid water-compatible packing of the necessary properties. Some work has been done on glass of controlled pore size, but with many natural polymers adsorption occurs and with proteins in particular the solutes have a tendency to be irreversibly bound to the glass surface. Modification of the packings by reaction with silylating reagents has led to vast improvements (16, 17) and there seems to be no reason why the microparticulate packings that have been so treated should not be very widely used for biochemical applications. In fact, human plasma proteins have been separated by μ-Bondagel-E columns (13), while chemically modified controlled pore glass (although of larger particle size) has been used for the separation of human serum components and dextrans (18).

The use of size-exclusion chromatography to separate compounds of intermediate molecular weight is a relatively new field. Separations have, however, been recently published, and it appears certain that for many of the polar non-volatile materials of molecular weight up to about 3000, size-exclusion will become the separation technique most widely used. As an example of what can be achieved figure 7.6 shows the separation of various hydrocarbons ranging in molecular weight from 506 to 72 (19). A more relevant separation, that of an epoxy resin, is shown in figure 7.7 (20).

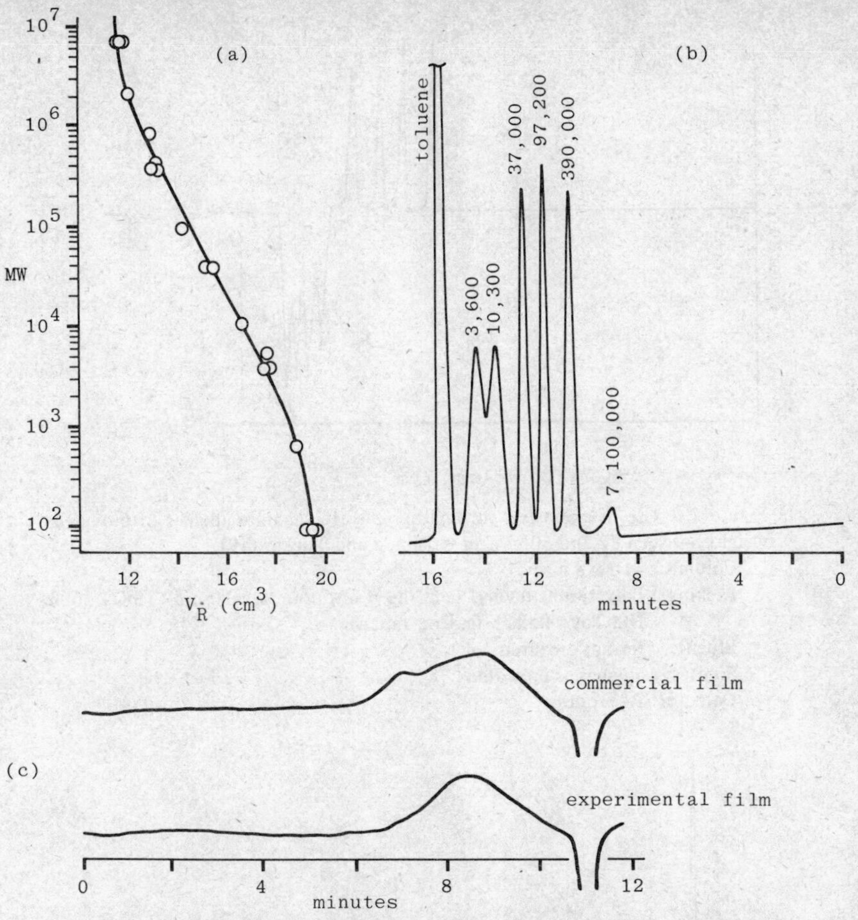

Figure 7.5. HPLC on column sets consisting of four columns of porous silica microspheres (12). (a) and (b) Calibration curve and chromatograms of Polystyrene standards (MW's indicated):

Column set:	150 mm PSM-300S + 100 mm PSM-800S + 100 mm PSM-1500S + 150 mm PSM-4000S, all 7.8 mm bore
Eluent:	tetrahydrofuran, 23°C
Detector:	UV, 254 nm, 0.1 AUFS

(c) Characterisation of cellulose films:

Column set:	100 mm x 7.8 mm of each of PSM-50S, PSM-800S, PSM-1500S and PSM-4000S supports silanised to reduce adsorption
Detector:	RI monitor, 5 x 10^{-5} RIU full scale
Eluent:	dimethylsulphoxide, 23°C.

It can be seen that size-exclusion chromatography has now become not only a very rapid technique for the determination of the molecular weight distribution of polymers but also a widely used method for the rapid and efficient separation of a large variety of chemical compounds. It seems likely that with the advent of chemically and mechanically stable size-exclusion packing materials based on silica this trend will continue until separation by size becomes one of the accepted techniques used for the separation of any complex mixture.

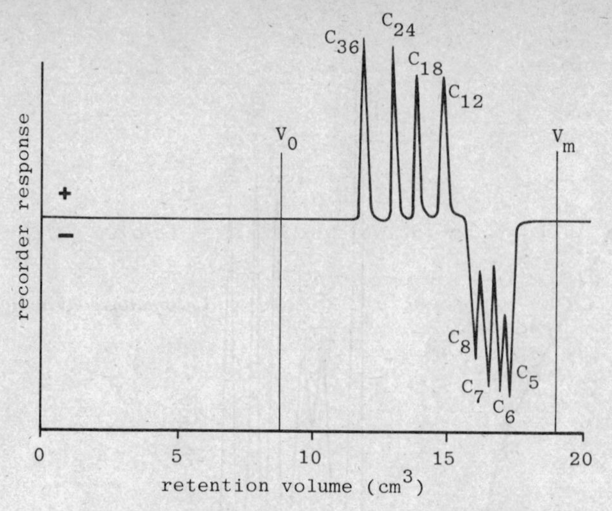

Figure 7.6. Use of exclusion chromatography to separate alkanes of molecular weight between 72 and 506 (after Krishnen and Tucker (19)).

Column: 610 x 8 mm

Packing: Polystyrene-divinyl benzene resin, pore size 10 nm (TSKG 2000 H8, Toyo Soda Mfg. Co., Japan)

Eluent: tetrahydrofuran

Solutes: alkanes as indicated

Detector: RI monitor

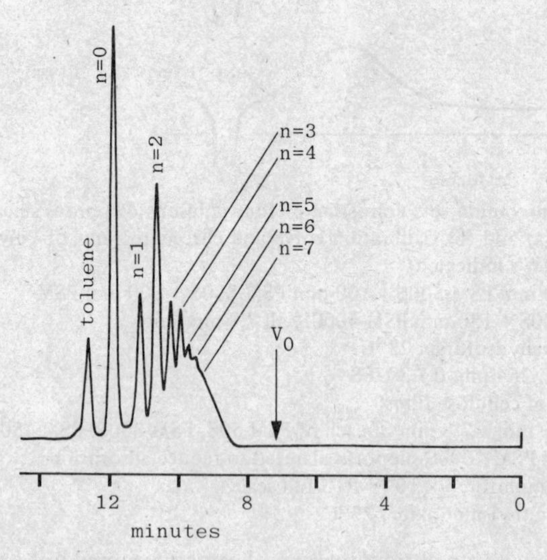

Figure 7.7. Separation of oligomers of "Epon 840" epoxy resin (after Kirkland and Antle (20)).

Column: 1000 x 6.2 mm

Packing: 8.4 μm porous silica microspheres PSM-40S, mean pore size 6 nm

Eluent: tetrahydrofuran, 23°C

Solutes: Oligomers of epoxy resin

Detector: UV photometer, 254 nm, 0.2 AUFS

REFERENCES

1. Porath, J. and Flodin, P., *Nature 183* (1959) 1657.
2. Porath, J., *J. Pure Appl. Chem. 6* (1963) 233.
3. Moore, J.C., *J. Polym. Sci. A–2* (1964) 835.
4. Grubisic, Z., Rempp, P. and Benoit, H., *J. Polym. Sci.* (1967) 753*B*.
5. Giddings, J.C., Kucera, E., Russell, C.P. and Meyers, M.B., *J. Phys. Chem. 72* (1968) 4397.
6. Casassa, E.F., *J. Phys. Chem. 75* (1971) 3929.
7. Van Krefeld, E. and Van den Hoed, N., *J. Chromatogr. 83* (1973) 111.
8. Yau, W.W., Kirkland, J.J., Bly, D.D. and Stoklosa, H.J., *J. Chromatogr. 125* (1976) 219.
9. Knox, J.H. and McLennan, F., *Chromatographia 10* (1977) 75.
10. Billingham, N.C., in *Practical High Performance Liquid Chromatography* (ed. Simpson, C.F.). Heyden and Son, London, 1976.
11. Scott, R.P.W. and Kucera, P., *J. Chromatogr. 125* (1976) 251.
12. Kirkland, J.J., *J. Chromatogr. 125* (1976) 231.
13. Mori, S., *J. Appl. Polym. Sci. 21* (1977) 1921.
14. Vivilecchia, R.V., Lightbody, B.G., Thimot, N.Z. and Quinn, H.M., *J. Chromatogr. Sci. 15* (1977) 424.
15. Alsop, R.M., Byrne, G.A., Done, J.N., Earl, I.E. and Gibbs, R., *Process Biochemistry 12* (1977) 15.
16. Regnier, F.E. and Noel, R., *J. Chromatogr. Sci. 14* (1976) 316.
17. Engelhardt, H. and Mathes, D., *J. Chromatogr. 142* (1977) 311.
18. Persiani, C., Cukor, P. and French, K., *J. Chromatogr. Sci., 14* (1976) 417.
19. Krishen, A. and Tucker, R.G., *Anal. Chem. 49* (1977) 898.
20. Kirkland, J.J. and Antle, P.E., *J. Chromatogr. Sci. 15* (1977) 137.

8. COLUMNS, CONNECTORS, INJECTORS

The column is the core of any chromatograph, and a primary aim in designing any chromatograph must be to ensure that the full potential of the column is realised in the recorder trace. The kind of performance that should be achieved is illustrated by the test chromatogram shown in figure 1.2. Unfortunately, it is only too easy to lose the resolution that can be generated by the column through badly designed or assembled injectors, detectors and connecting pieces. Much thought has been given to the design of these critical components immediately adjacent to the column, and it is the purpose of this chapter to outline some of the main considerations in this regard.

A modern high-performance LC column will typically have a bore of 5 mm or more and a length of 100 mm or more. A column of 5 mm-bore, 125 mm long, will require about 1.6 g of silica for packing, and will contain about 1.9 cm^3 of eluent, this being about 75% of the total volume of the empty column. If well packed with 5 μm particles such a column will develop around 10,000 theoretical plates when operated at the minimum in the (h, ν) curve. The peak volume (i.e. the volume of eluent within which the peak base elutes) should then be about 75 mm^3 for an unretained solute peak (equation 2.7c). For the full potential of the column to be realised this width must not be broadened significantly by the operations of injection and detection and by passage of the solute through any intermediate tubing and connectors.

8.1 KEY IDEAS

The relationship that states how the dispersive effects of these different parts of the equipment are combined is given by

$$\begin{aligned} w_{total}^2 &= w_v^2 + w_{inj}^2 + w_{con}^2 + w_{det}^2 \\ &= w_v^2 + w_{app}^2 \end{aligned} \tag{8.1}$$

where w_{total} is the total peak width measured by the recorder, w_v is the peak width produced by the column alone, w_{inj} the peak width produced by the injection alone, and w_{con} and w_{det} the peak widths produced by the connections and detector. The combination of the last three terms, w_{app}, is the peak volume due to the equipment other than the column, and can be obtained experimentally by connecting the injector directly to the detector omitting the column, or rather by replacing the column by a fitting where the packed section is omitted. For the best systems, w_{app} directly determined in this way is between 30 and 40 mm^3. Suppose then that w_v = 75 mm^3, we can then calculate w_{total}:

$$w^2_{total} = 75^2 + 35^2 = 6850\ mm^6$$

$$w_{total} = 83\ mm^3 \qquad (8.2)$$

It is notable that the additional peak spreading due to the extra column dispersion processes is only about 10% when the peak width due to these processes alone is nearly half that due to the column.

While this result seems fairly favourable it is in fact all too easy to introduce massive additional peak spreading by poor design of the critical components, and indeed by poor design of the column itself.

8.2 COLUMN BORE

Experience has shown that with microparticles higher overall efficiencies are generally obtained with wider columns. There are three independent reasons for this:

(a) *Peak Volume*. As shown in equation 2.7c, the peak volume arising from a point injection is given by

$$w_v = 4V_R/N^{1/2} \qquad (8.3)$$

where V_R is the retention volume and N the number of theoretical plates to which the column is equivalent. As noted above, the spreading due to extra column effects in the best HPLC systems amounts to about 35 mm^3. If it is agreed that this should cause less than 10% excess broadening in the worst case of an unretained solute peak, then w_v must be greater than 75 mm^3. For a 10,000-plate column equation 8.3 then gives $V_R \geqslant 1900\ mm^3$. Since the eluent volume in the column represents only 75% of the volume of the empty column, the minimum volume of the empty column is 2500 mm^3. For a column 125 mm long this means that the bore must be at least 5 mm. Narrower columns will fail to give their full performance when the additional dispersion from the ancillary equipment is added. This is dramatically demonstrated by figure 8.1, which shows the apparent plate count observed for columns of different bore. The columns are assumed to be 125 mm long and to be equivalent on their own to 10,000 theoretical plates: the extra column spreading is taken as 35 mm^3 for all the columns.

In general, for columns equivalent to N theoretical plates the calculation just given would indicate a minimum column volume of approximately $25N^{1/2}$ mm^3.

(b) *"Infinite Diameter Effects"*. Centrally injected samples of solute spread across a column packed with fine particles quite slowly during elution, so that it is quite possible for a centrally injected sample to traverse the entire column without an appreciable fraction of the solute reaching the walls. The theory of the effect has been considered in detail elsewhere (1,2), but for the present purpose it is simply noted that with a 125 mm column packed with 5 μm particles and operated at the optimum velocity, the radial spreading from a point injection amounts to no more than 1.3 mm. If desired, it is therefore quite possible to design HPLC columns so that the injected solute completely avoids the wall regions where the regularity of packing might be disturbed. To make practical

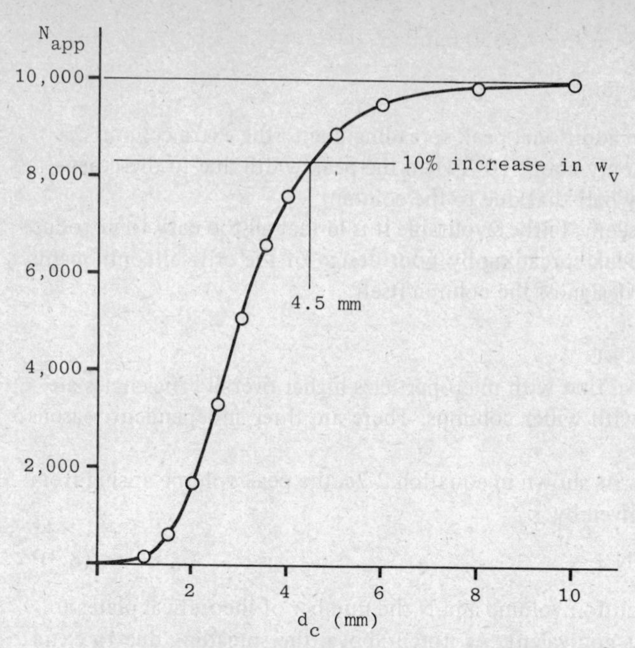

Figure 8.1. Effect of extra column dispersion of 35 mm^3 on the apparent plate number for 125 mm columns of different bores that would be equivalent to 10,000 theoretical plates when tested with a perfect detector and an unretained solute.

HPLC columns that behave as if they were of "infinite diameter" in this sense, one has to allow a fair tolerance to cope with the initial size of the injection sphere, and with any eccentricity of the injection arising, for example, from the use of an angled syringe needle. A reasonable allowance for all effects is 1 or 2 mm on the radius of the column or 2 to 4 mm on the bore. The minimum safe bore if one wishes to operate columns of microparticles in the infinite diameter mode is therefore 5 to 7 mm.

(c) *The "Extent of the Wall Region"*. Whether or not it is important to avoid the wall regions of the column depends upon how well or how badly the column is packed in the region of the walls. The evidence available (2) suggests that there is normally a badly packed region close to the walls where the dispersion of the solute is much greater than in the core of the column. This layer appears to extend about 30 particle diameters inwards from the wall. For a column of 5 mm bore containing 20 μm particles the fraction of the column cross-section taken up by this annular wall layer is about 40% and is therefore very significant. However, if the column is packed with 5 μm particles the extent of the wall region is only about 12%, and while it will be rather important to avoid the wall region in the first case it will be less important in the second. The conclusion one reaches is that with a sufficiently wide column the extent of wall region can be so small that it becomes unimportant whether solute reaches the walls or not. The critical column diameter for 5 μm particles is again about 5 mm.

These three separate considerations all suggest independently that the minimum column bore, if one is to obtain the highest performance with 5 μm particles, is in the region of 5 mm, and that one may expect some further improvement from going up to 7 mm.

8.3 CONNECTING TUBING AND CONNECTIONS

The resolution of solutes produced by a column can easily be lost if connecting tubing is too long, of too large bore, or if connections contain dead volumes.

The volume dispersion, w_{tube}, produced by flow through a smooth-walled circular tube is given by Taylor's equation (3). For passage of an element of solution along an open tube of diameter d and length L, this is given by

$$w_{tube} = 0.36\, d^2 \left(\frac{Lf_v}{D_m}\right)^{1/2} \qquad (8.4)$$

where f_v is the volume flow rate and D_m the diffusion coefficient of solute in eluent. If 30 mm^3 is taken as the maximum allowable value for w_{tube}, and if $f_v = 20\ mm^3\ s^{-1}$ and $D_m = 10^{-9} m^2 s^{-1}$ (water) we obtain the following maximum allowable tube lengths:

Bore		*Maximum*
(mm)	(in)	*length* (mm)
0.125	0.005	1500
0.15	0.006	800
0.25	0.01	100
0.5	0.02	6

In practice somewhat greater lengths can be allowed, because the flow will not be truly laminar in the relatively rough-walled tubing that is generally used. Nevertheless it is evident that in making any connections between the injector and detector the tubing bore must not be more than 0.25 mm (0.010 in), and that still finer tubing would be preferable, although one must then take account of the likely pressure drop and possible blockage of the tubing.

Connections between tubing must be made with care, and in all cases should be made by zero-dead-volume fittings in which the connecting tubes are butted together. Normal 1/16 in Swagelok connections, for example, must be drilled out to remove the shoulder in the central section.

8.4 COLUMN DESIGN

Columns for HPLC should evidently be short and fat, unlike those for GC, which are long and thin. Typical sizes for the packed bed in an HPLC column are therefore 100–250 mm in length and 5–8 mm in bore. Because of dead volume and wall effects it is desirable to inject the sample as near to the top of the bed as possible and to make the injection in an axially symmetrical manner so that maximum use is made of the core of the column. Similar considerations apply in reverse at the column outlet.

(a) *The Column Outlet Fitting*

The column packing must be held in the column by a packing retainer at the outlet end. This is generally either a fine porosity metal frit or a disk of fine stainless steel gauze or cloth. These must be able to retain particles larger than one-third of the nominal mean particle size. The retainer can be fitted either into the end of the column or into the coupling that fits the end of the column. Four possible arrangements are illustrated in figure 8.2. There is no detectable difference in chromatographic performance between these four designs but there are some relevant practical points to be noted.

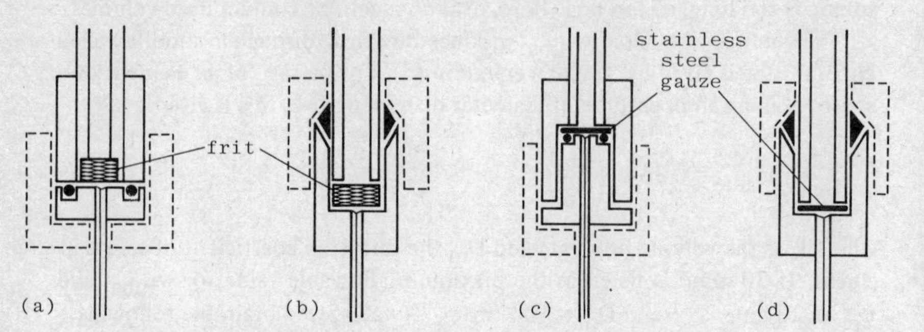

Figure 8.2. Outline designs for column outlet fittings. Retaining nuts shown by broken lines.

Frits are more readily blocked than wire cloth gauzes, and are generally difficult to clean. When mounted inside the column, they have to be inserted carefully to avoid deformation and closure. For mounting within the column the gauze disk therefore is preferable. The systems shown in figures 8.2b and d are based upon commercial tube fittings, but these normally have to be modified by drilling to eliminate dead volumes. In fittings of the kind shown the tube is pressed home onto the metal frit or screen by the act of tightening the coupling. Because of the problem of tolerance in the dimensions of the fittings it is not subsequently possible to change the coupling; the same column and end coupling must be used together from that time onwards. The systems shown in figures 8.2a and c do not suffer from this problem. Considering all the above factors, that shown in figure 8.2c is the most satisfactory system for long term use and re-use. The O-ring may be of PTFE or a suitable inert elastomer.

(b) *Top Column Fitting*

Again there are a number of more or less satisfactory systems. Many manufacturers supply columns that are terminated in the same way at both ends. When supplied these columns must be entered and exited by 1/16 in narrow-bore tubing, and this does not allow on-column injection. Another disadvantage is that the packing may contract in use due to settling and leave a void at the inlet end. Such a void is undetectable except from the serious loss of column performance that it causes. We believe that it is important that the inlet end of the column should be easy to dismantle for inspection, cleaning and topping up if necessary.

In the past injection by syringe was sometimes made directly into the column packing itself. While this can give excellent results for the first few injections, the top of the packing is inevitably disturbed, and to maintain efficiency successive injections have to be made progressively deeper into the bed. The best way to avoid this problem but still to maintain high efficiency is to protect the top of the column packing by a gauze screen. Injection is then made into a shallow non-dispersive section just above the screen. Two convenient systems are shown in figures 8.3a and b. The system shown in figure 8.3a is based upon a commercial tube fitting. The gauze disk is simply placed upon the top of the packing and topped by a 5 mm layer of 100-mesh (200 μm) glass beads into which injection is made. Figure 8.3b shows a more elaborate system, in which the gauze screen is mounted on a shelf and held in position with a PTFE or metal ring. The hole in the ring is filled with glass beads. This system has the advantage that there is no possible leakage of eluent round the edge of the disk, and that the disk is maintained strictly horizontal. The seal to the injector is made by a PTFE or elastomer O-ring. The use of a swaged connection at the column head, especially when connected direct to the injector, is unsatisfactory in the long term because it is very easy to damage the high-pressure seal with particles of dust or column packing. This is likely to occur after 10 to 20 dismantlings, whereupon both the column and the injector may have to be discarded or at least refaced. For the system sealed by an O-ring only the O-ring need be replaced.

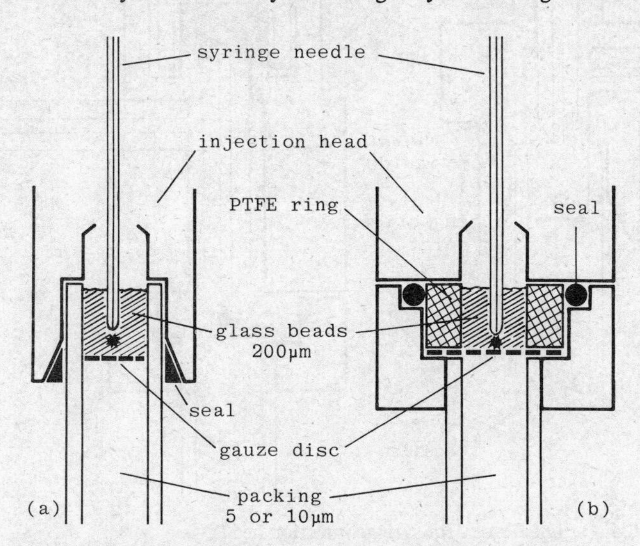

Figure 8.3. Outline designs for column inlet fittings, showing syringe needle in the injection position. Fixing nuts are omitted.

8.5 SYRINGE INJECTION

Injection by micro-syringe (1 to 10 mm^3) is widely used, although valve injection is becoming increasingly popular, especially in routine applications.

Injection by syringe suffers from a number of problems but at its best provides the highest efficiency of any injection method. To make the maximum use of the core of the column, the injection should be made sharply at the points marked * in figure 8.3a and b. In this way the injection is made into a curtain flow of eluent, so that the sample is carried rapidly downwards with little or no radial dispersion before it reaches the packing proper. Numerous syringe injector designs are available. In virtually all of them the syringe needle passes through a self-sealing rubber septum (normally 1/4 in diameter and 1/8 in thick) which is rigidly held to prevent extrusion. Two designs of injector are shown in figure 8.4: that of 8.4a is based upon a modified commercial tube fitting while that of 8.4b is designed without reference to such constraints.

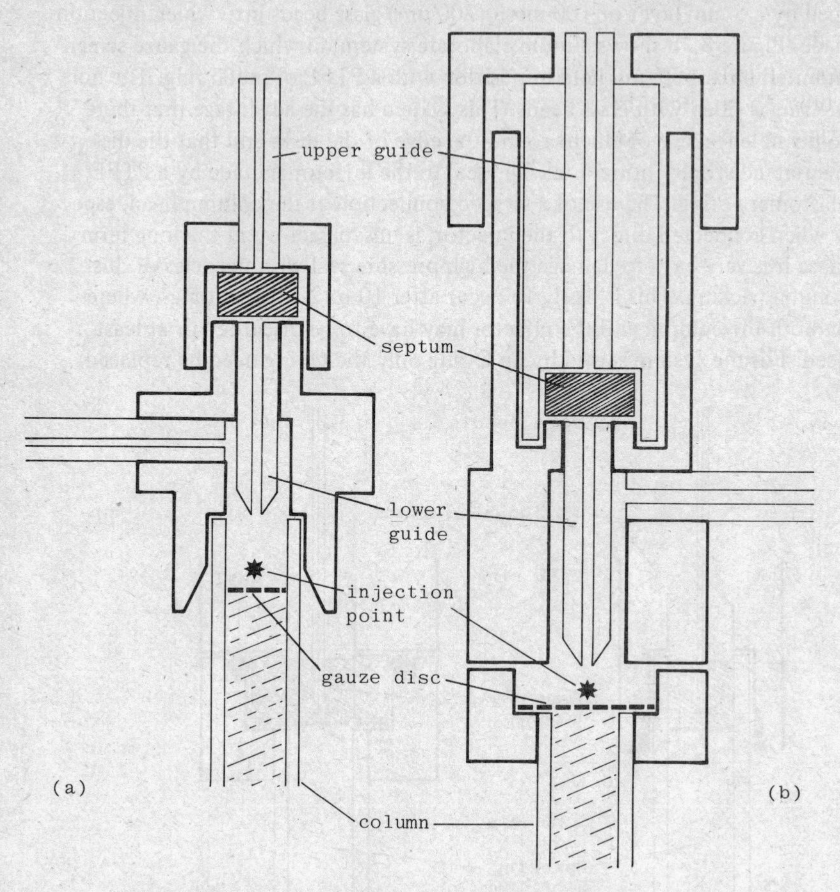

Figure 8.4. Outline designs for syringe injection heads.

An important feature of any good injector is that the septum should be as near to the bottom of the injector as possible, that it should be easy to remove, and that the needle should be guided both above and below the septum. The guide above the septum ensures that the needle remains unbent when the septum is pierced, while the lower guide is necessary to ensure central injection into the column. The total length of the injector should be such that when the syringe with a needle of the correct length is pushed home to its full extent the tip of the needle is about 2 mm above the gauze disk.

A major problem with syringe injection is blockage of the needle by pieces of septum rubber. This arises because, under high compression, the normal sharp-ended syringe needle cuts of small portions of rubber at each penetration of the septum. These pieces of rubber then tend to enter the needle. With a glass syringe, if one tries to force the rubber out of the needle by pressing down the plunger of the syringe, the barrel will be split. This particular problem is almost entirely eliminated if one uses septa that have been pre-drilled with a 0.5 or 1 mm hole, and syringe needles with smooth domed ends that cannot easily cut the septum rubber.

Inevitably, even when using domed needles and pre-drilled septa, small pieces of septum rubber do become detached and are pushed down the lower guide tube. These pieces come to rest in the glass bead layer and must be removed from time to time. It is therefore important that the column can be easily and repeatedly disconnected. The system shown in figures 8.3b and 8.4b allows this to be done more easily and more often than that shown in figures 8.3a and 8.4a.

Syringe injection can be used up to pressures of about 100 bar (1500 psi) and gives a reproducibility for repeat injections of the order of 5% on absolute quantity, or 1% relative to an internal standard.

The main cause of irreproducibility in syringe injection arises from leakage past the plunger. With water at 1000 psi a typical leakage rate from a 10 mm^3 micro-syringe with steel plunger and glass barrel is 0.1 mm^3 s^{-1}.

8.6 INJECTION VALVES

The simplest injection valves are six-way valves, which work on the principle shown in figure 8.5. Here the injection loop is shown external to the valve, and this arrangement is convenient for injection volumes of 20 mm^3 and above.

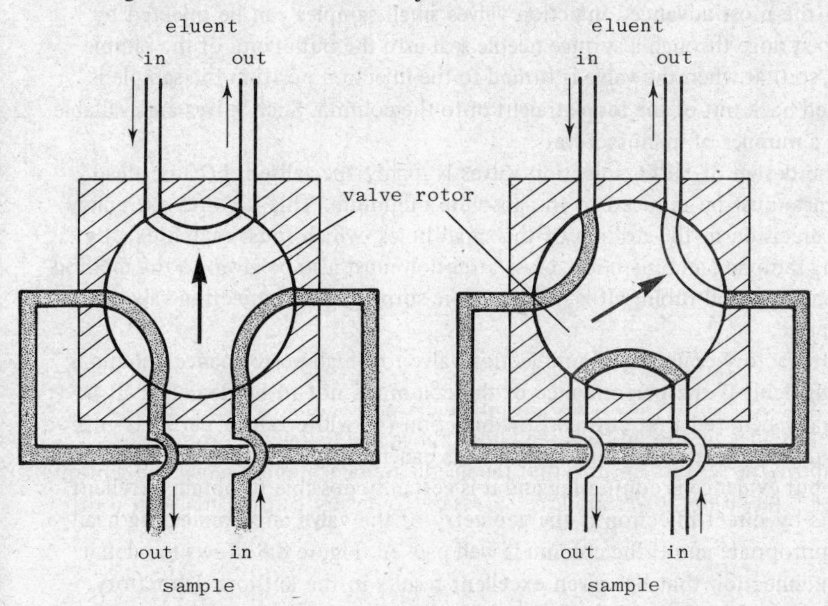

Figure 8.5. Principle of operation of a 6-port injection valve with external sample loop.

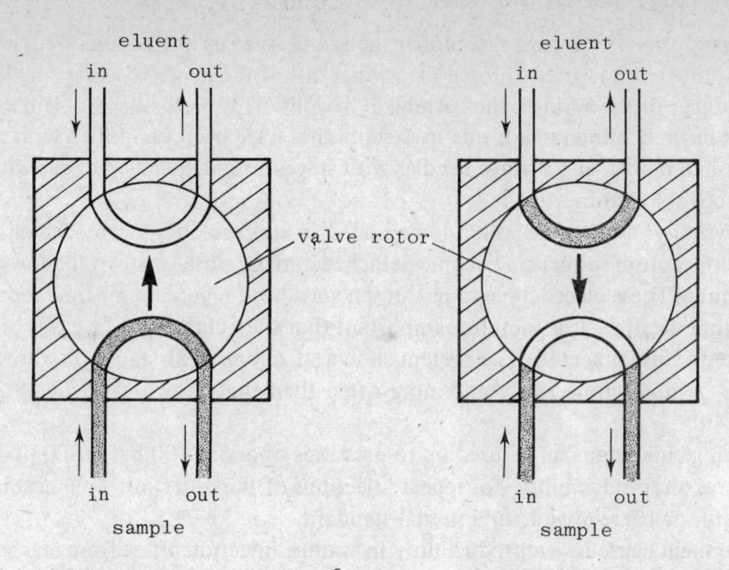

Figure 8.6. Principle of operation of a 4-port injection valve with internal sample loop.

For smaller injections the sample volume is within the central rotor of the valve. In this case the simpler arrangement shown in figure 8.6 can be used, which employs a four-way valve.

In the most advanced injection valves small samples can be injected by micro-syringe through a syringe needle seal into the outlet end of the sample loop, so that when the valve is turned to the injection position the sample is flushed back out of the loop straight onto the column. Such valves are available from a number of manufacturers.

The design of HPLC injection valves is highly specialised, because dead volumes must be reduced to the absolute minimum. This requires extremely high precision in the drilling of the small holes, which must match exactly during sampling and injection; close attention must also be given to the method of connecting all tubing. It is not therefore surprising that injection valves are expensive.

Satisfactory coupling of an injection valve to a high-performance column is also difficult if the performance of the column is not to be degraded. It is generally believed that curtain-flow injection (4) will produce better results than direct injection because wall effects can thereby be minimised (figure 8.7), but evidence is conflicting and it is certainly possible to obtain excellent results by direct injection if the geometry of the valve and connecting head are appropriate and if the column is well packed. Figure 8.8 shows the detail of a connection that has given excellent results in the authors' laboratory.

The key factors for the success of this design appears to be (a) that the bore of the 1/16 in connecting tube must not exceed 0.25 mm (0.010 in); (b) the end of the 1/16-in connecting tube must be flush with the end of the

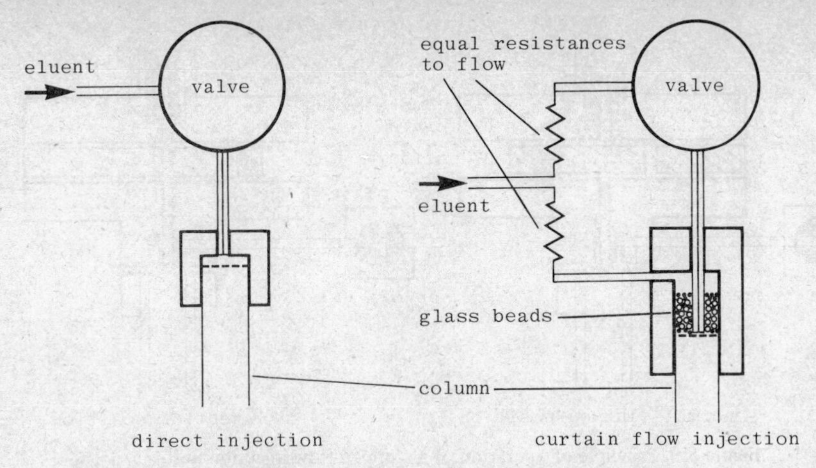

Figure 8.7. Methods of coupling injection valves to columns.

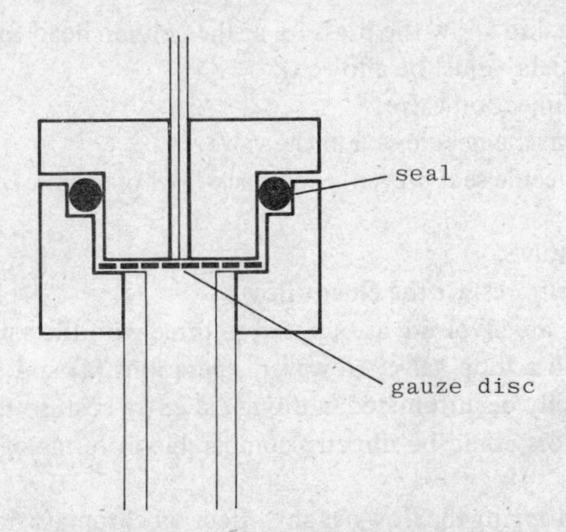

Figure 8.8. Detail of direct injection head for use with valve injectors.

connecting plug; (c) the connecting plug must press home onto the shelf that supports the gauze disk, (d) there must be no gap whatever in the column packing under the gauze disk. With this arrangement a column 125 mm × 5 mm, which gave 8000 theoretical plates using syringe injection, gave 7000 plates using injection by a valve. However, the method of packing the column is undoubtedly the most critical factor, and this is discussed in chapter 12.

8.7 STOPPED-FLOW INJECTORS

The stopped-flow injector enables one to obtain the advantages of syringe injection without the disadvantages of septa. Several models are commercially available and the essential principle is illustrated in figure 8.9. The injector proper consists of a valve that can be opened at ambient pressure to allow the injection of sample by a syringe. This is done after stopping the flow of eluent from the high-pressure pump by an auxiliary on/off valve. The sequence of operations for stopped-flow injection is as follows:

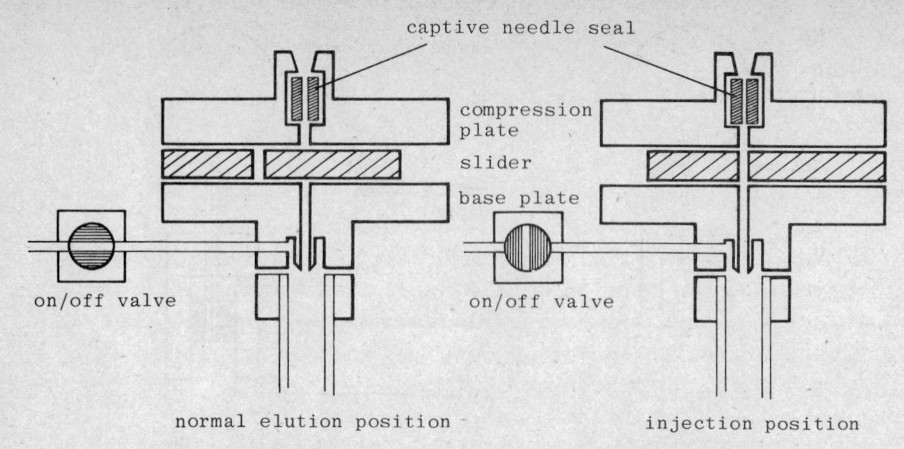

Figure 8.9. Principle of operation of a stopped-flow injection unit.

(a) close the auxiliary valve and allow the pressure at the column head to fall to ambient (roughly 10 s delay must be allowed);
(b) open the stopped-flow injection valve;
(c) insert the needle through the needle seal in the valve;
(d) inject the sample (the needle seal prevents any back-flow of sample);
(e) withdraw the syringe;
(f) close the stopped-flow valve;
(g) open the auxiliary valve to restart the eluent flow.

Seven independent operations are involved, as opposed to three with the simple syringe injection and five with a loop valve. However, operations (a) and (b) along with (f) and (g) can readily be automated, although at extra cost, so that stopped-flow and valve injectors could be directly comparable in terms of operational convenience.

There is no evidence that stopping the flow has any effect on chromatographic performance or quantitative accuracy, but some detection systems are sufficiently flow sensitive that an unduly large injection peak may result from stopping and starting the flow.

The main advantages of the stopped-flow system are that it enables the sample to be injected in the best possible way onto the head of the column, it allows economy of use of scarce samples, and it can be used when the column operating pressure is high.

8.8 SELECTION OF METHOD OF INJECTION

Selection of the method of injection depends upon individual requirements and probably no single injector will be ideal for all purposes.

(a) *Cost*. Valve injectors are relatively expensive but running costs are low. Syringe injectors are relatively cheap but the running costs can be high if syringes are frequently broken. However, as pointed out above, the use of domed needles largely prevents this. Stopped-flow injectors are intermediate in cost, depending upon their sophistication.

(b) *Ease of Operation*. Valve injection requires the least operator skill and so is the system of choice for routine applications where operators are likely to be relatively unskilled. On the other hand, skilled technicians and research workers find syringe/septum injection convenient and economical.

(c) *Reproducibility*. Syringe injection will give reproducibility of around 5% on an absolute basis and around 1% on a relative basis. Valve injection is more precise, with a reproducibility of 1% or better on both bases.

(d) *Sample Size*. Syringe injection, whether using a septum system, a stopped-flow system or one of the sophisticated injection valves, allows samples of any size to be injected up to some instrumental maximum. Loop injectors, where the loop must be completely filled, are, of course, restricted to the volume of the loop. This may be a disadvantage in a research environment.

(e) *Economy of Sample Solution*. Syringe injection requires only the volume of sample that is actually injected into the column. It is therefore highly economical of scarce sample solutions. Valve injection, where a loop has to be completely filled, generally requires about 100 mm^3 of sample per injection. This may be more than one can afford of a precious sample.

(f) *Operating Pressure*. Valve and stopped-flow injectors are generally rated up to at least 330 bar (5000 psi). Syringe/septum systems cannot easily be used at pressures above 100 bar (1500 psi), although glass syringes will stand considerably higher pressures and the type B syringes with the plunger in the needle will withstand very high pressures. The septum is the limiting factor, not the syringe.

8.9 SUMMARY

Recommendations regarding column and injection design and operation may be summarised as follows:

(a) The column bore should not be less than 5 mm for columns 125 mm in length; the total volume of a 10,000-plate column must be at least 2500 mm^3. For columns equivalent to N plates the minimum column volume (in mm^3) is given approximately by $V_{col} \geqslant 25N^{1/2}$.

(b) Injection should be made as close to the top of the column packing as possible and should be axially symmetrical.

(c) There must be no voids in the system between the injector and detector. Voids in the column packing are especially serious.

(d) Connecting tubing should not be more than 0.25 mm in bore (0.010 in), and the length should be kept short, say not more than 100 or 200 mm.

(e) Connections between narrow-bore tubing should be made by butting the two ends together.

(f) Detection cells should be as small as possible, and never larger than 1 mm bore and 10 mm long (typical dimensions for a photometer cell).

REFERENCES

1. Horne, D.S., Knox, J.H. and McLaren, L., *Separ. Sci. 1* (1966) 531.
2. Knox, J.H., Laird, G.R. and Raven, P.A., *J. Chromatogr. 122* (1976) 129.
3. Taylor, Sir G., *Proc. Roy. Soc. A219* (1953) 186.
4. Webber, T.J.N. and McKerrell, E.H., *J. Chromatogr. 122* (1976) 243.

9. PUMPS AND GRADIENT SYSTEMS

The function of the pump in a high-performance liquid chromatograph is to induce a flow of eluent to the column at a controlled rate, f_V, and at an elevated pressure, ΔP. These two variables are related by equation 9.1 through the downstream flow resistance, R, which arises in the column and any other restrictions to flow between the pump and outlet from the detector.

$$\Delta P = f_V R \tag{9.1}$$

HPLC pumps are of two main types. The first type comprises those that primarily control the pressure, ΔP, so that the flow rate f_V is determined by the resistance, R. Coil pumps and pneumatic intensifiers are of this type. The second type comprises those that primarily control the flow rate, f_V, so that the pressure drop, ΔP, is determined by the resistance, R. All pure mechanically driven pumps are of this type.

Typical flow rates in analytical HPLC are in the region of 0.5 to 2 cm^3 min^{-1}, and for adequate flexibility pumps should be capable of delivering eluent over a range of flow rates from 0.1 to 10 or 20 cm^3 min^{-1}. Typical operating pressures range from 5 to 100 bar (75 to 1500 psi) and an acceptable pump having something in hand will be capable of delivering eluent against a column back pressure of not less than 300 bar (4500 psi).

9.1 DESIGN CRITERIA FOR HPLC PUMPS

Satisfactory pumping systems for HPLC must meet a number of design criteria, of which the most important are the following:

(a) *Materials*. Because HPLC pumps must handle a wide range of working fluids, parts exposed to the eluent are customarily constructed of glass or ceramic, austenitic stainless steel (AISI 316 grade or equivalent), PTFE, or one of a limited range of synthetic elastomers that must be selected according to the type of eluent used. Wetted components are joined together by mechanical joints in the same or compatible materials, by stainless steel welding, or by brazing with gold-nickel alloys.

(b) *Flow Constancy*. The fluid flow rate from the pump must be both smooth and uniform to within about 1% at all usable flow rates for two important reasons. Firstly, quantitative accuracy of this order will only be obtainable if flow is reproducibly precise; secondly, the flow sensitivity of many detector systems will lead to baseline noise if flow fluctuations occur. Smooth flow with pumps having large-area pistons (intensifier and large syringe pumps) can be achieved only if their sealing systems have uniform frictional properties and minimal stick-slip characteristics. The seal requirements for pumps with small-area pistons, such as small-volume reciprocating pumps, are less severe, but such pumps may give a non-uniform output that requires a pulse damping or smoothing system, unless they are of the dual-head type driven by specially designed cams or gears.

(c) *Access*. Ready access to the flow adjustment mechanism and to the high-pressure sliding seals and non-return valves is essential. These components are critical to the performance of any pumping system and they are also nearly always those parts that wear most rapidly. Good access to the low-pressure side of any high-pressure seal housing is also important, for this is where deposits of solids from slight leakage of eluent (e.g. salts from buffer solutions) may build up sufficiently to wear away soft seals or even metal parts.

(d) *Pressure Requirement*. Any HPLC pump must be able to pump against the inlet back pressure of the chromatographic column when operating at the appropriate flow rate. In the current state of the art, using 5–10 μm particles and short columns (50–250 mm) with bores of between 5 and 10 mm, column back pressures under normal operating conditions range from about 5 to 100 bar (75 to 1500 psi). The very high operating pressure limits quoted for most purpose-designed HPLC pumps are therefore very rarely used at the present time, but this is unlikely to be a permanent feature of HPLC as chromatographers strive for shorter separation times and greater resolving power. These improvements will only be achieved by using smaller particles, or longer columns, or both. Consequently higher pressures will again be in demand.

9.2 TYPES OF PUMP

The five types of pump listed in chapter 1 are discussed in detail in the following sub-sections:

(a) *Coil Pumps*

These are in principle the simplest possible source of pressurised liquid, and have the great advantage, having no moving parts, that they cannot suffer from flow anomalies due to friction nor from seal failure. As shown in figure 9.1, the eluent liquid is stored in a coil of stainless steel tubing having a volume of 200 to 400 cm^3. The coil is fitted with valves at each end so that it can be independently connected to the high-pressure air cylinder or to a vent line at the top end, and to the column or to a filling reservoir at the lower end. These valves must, however, be interlocked in such a way that valve A can be opened to the high-pressure gas line only when B is open to the column, and before the coil is refilled the high-pressure gas must be vented by opening A to waste before B is opened to the reservoir. Coil pumps are trouble-free in operation and refilling with the same eluent is straightforward, but when the eluent is to be changed the holding coil must be emptied and then washed several times with the new operating solvent. Operator time and a considerable volume of eluent liquid is thereby wasted.

The liquid pressure is limited to that obtainable from gas cylinders, but with current column technology this is not a serious disadvantage, except that when working at pressures at or above 100 bar (1500 psi) only about half or less of the contents of a full gas cylinder can be used. The use of high-

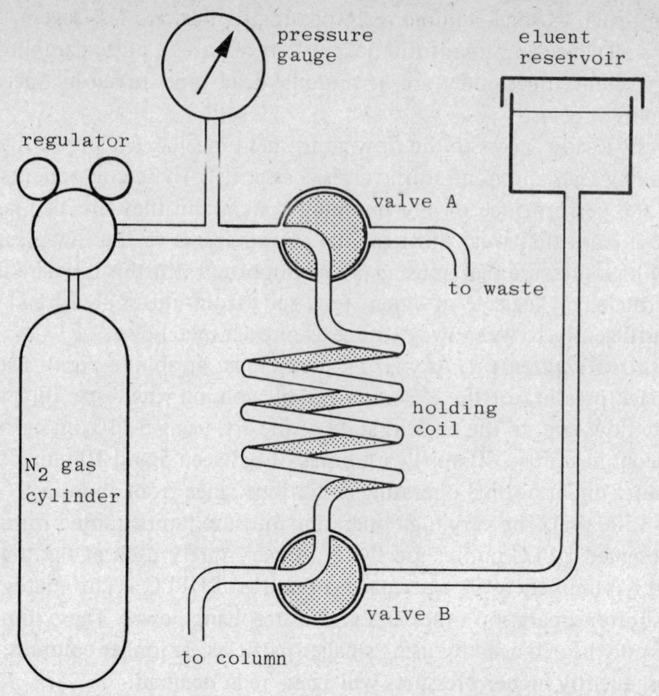

Figure 9.1. Coil pump operated by high-pressure gas. Coil volume about 200 cm^3.

pressure gas does present safety problems, since high-pressure gases are much more dangerous than liquids under similar conditions. All valving and tubing must be made to a high standard, and any interlock system must be robust and accident-proof. As a result there is little difference in price between a safe coil system, a well-damped single-piston reciprocating pump, or a simple pneumatic intensifier.

(b) *Pneumatic and Hydraulic Pressure Intensifiers*

A gas pressure intensifier is illustrated diagramatically in figure 9.2; it develops a fluid output pressure that is proportional to the energising pressure of gas exerted on the face of a large-diameter piston A. The proportionality factor is the ratio of the cross-sectional area of the low-pressure piston A to that of the high-pressure piston B and is generally between 9 and 50 for pumps used in HPLC. Because of the pressure amplification it is possible to use conventional low-pressure industrial pneumatic controls (10 bar maximum pressure) to govern the refill, start and return operations.

For HPLC the basic pump would normally be fitted with inlet and outlet non-return valves, and the refill stroke would be activated by exhausting air pressure from the drive side and pressurising the reverse side of the piston. Refill may be rapid, and this high inlet flow rate imposes constraints on the type of in-line eluent filter that can be used. In our experience a 5 cm diameter disk of 400-mesh (37 μm porosity) stainless steel gauze suitably mounted is effective for a pump of 20 to 60 cm^3 displacement and does not induce excessive inlet flow resistance.

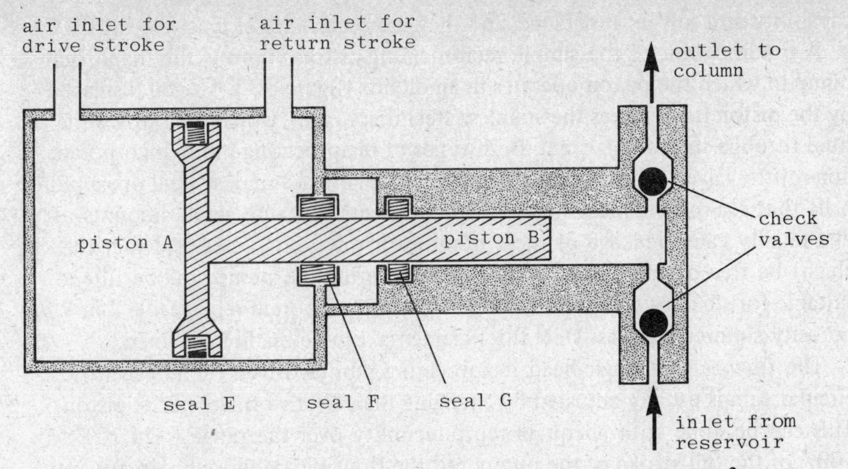

Figure 9.2. Pneumatic pressure intensifier, displacement 2 to 60 cm^3.

Intensifier pumps have sliding seals (E, F, G) in both the low- and high-pressure chambers. Since the only forces controlling the motion of the dual-piston assembly are sliding friction and the balance between the energising and the counteracting hydraulic forces, the frictional and stick-slip characteristics of the seals are critical. These problems have been largely solved in commercial pumps by special lubrication in the low-pressure chamber and the use of spring-loaded PTFE seals in the high-pressure chamber. Because the most widely used intensifier pumps are modifications of commercial intensifiers used for pressurising hydraulic oil, they do possess some infelicities of design when applied to HPLC. The most serious of these is difficult access to the low-pressure side of the high-pressure seal (G), which is normally contained within an aluminium alloy casting. This effectively prevents inspection for possible build-up of solid deposits that may ultimately damage all the seals in that part of the pump and can even cause it to seize solid. This problem can be avoided by routine washing of that part of the piston outside the high-pressure seal.

Nevertheless, under most conditions, and certainly with completely volatile eluents, intensifier pumps are very reliable; the ease with which the flow can be controlled and instantly stopped or started commends this type of pump to the routine user. A flow monitor is a useful addition to such pumps, for it enables the user to set the pressure to give the flow desired.

(c) *Single-Piston Reciprocating Pumps*

Single-piston reciprocating pumps, in which a piston of small area directly displaces the fluid being pumped, have long been used in amino-acid analysers, which may be considered the forerunners of modern HPLC equipment. They might therefore have had pride of place amongst the various pumping systems used in the development of HPLC. That this is not the case is possibly because the technology of amino-acid analysers had apparently come to a standstill when HPLC was being developed. Nevertheless several reciprocating pumps are now available that are highly reliable in respect of constancy of flow, pressure capability and solvent resistance, and they are used fairly extensively in HPLC.

A modification of the simple reciprocating-piston pump is the diaphragm pump in which the piston operates in an oil box (figure 9.4). The oil displaced by the piston itself flexes the stainless steel diaphragm, which then drives the fluid through the valve system. Both types of reciprocating pump incorporate non-return valves, which generally consist of spherical stainless steel or sapphire balls that seal on stainless steel or sapphire seats. Because small amounts of dirt readily cause leakage of these non-return valves, it is essential that the eluent be free of suspended solids before entering the pump. In-line filters suitable for small-swept-volume pumps, which incorporate replaceable 2 or 7 μm porosity sintered stainless steel filter elements, can be readily obtained.

The flow rate in single-head reciprocating pumps driven by an eccentric circular cam is usually adjusted by changing the effective travel of the piston. This can be done with adequate reproducibility over the range from 10% to 100% of the full stroke of the pump, but the flow inevitably pulsates, for even when working at full stroke only 50% of the total cam cycle is used for eluent pressurisation, the other 50% being used to refill the pump chamber. Pulse damping is therefore essential if the smooth flow required in HPLC is to be achieved. For relatively low pressures an air-filled snubber may be used; this is conveniently made from a 500 mm length of 5-mm-bore stainless steel tube capped at one end. The air in the tube is compressed as the pressure increases during the forward stroke and expands during the refill stroke. For higher pressures one can use either an hydraulic snubber, which may be spring loaded, or the hydraulic analogue of the capacity-resistance network used for electrical smoothing. This latter can be constructed from a sequence of Bourdon tubes in series with narrow-bore connecting tubes.

Although pulse dampers can be extremely effective in smoothing the pulsations due to the pump, they nearly always introduce additional fluid-filled chambers, in many cases poorly swept, between the pump and the injector. Change of eluent may then require considerable volumes to be passed to ensure that the solvent liquid reaching the column has the desired composition. Continuous change of eluent composition initiated on the inlet side of the pump is not practicable with most pulse-damping systems.

Accessibility of the seals and of the low-pressure side of the piston is generally excellent in simple reciprocating pumps, both for cleaning and replacement, but care is necessary to ensure that delicate and highly polished ceramic or sapphire pistons are not damaged when seals are replaced.

The most sophisticated single-piston reciprocating system, employing a variable stroke length to change the flow, uses an oil-filled chamber not only to damp pulsation but also to monitor and control the flow rate. It works on the following principle. During the pump refill stroke the damper relaxes and the pressure change in the damper, dp, is given by $dp = dV/\kappa V$ where κ is the compressibility of the oil, V the volume of the oil and dV the volume change. Thus the flow rate $f_V = dV/dt$ is given by $\kappa V\,(dp/dt)$. Since the pressure drop during this refill part of the cycle is small, the average dV/dt measured during this part of the stroke is effectively the same as the mean dV/dt over the whole

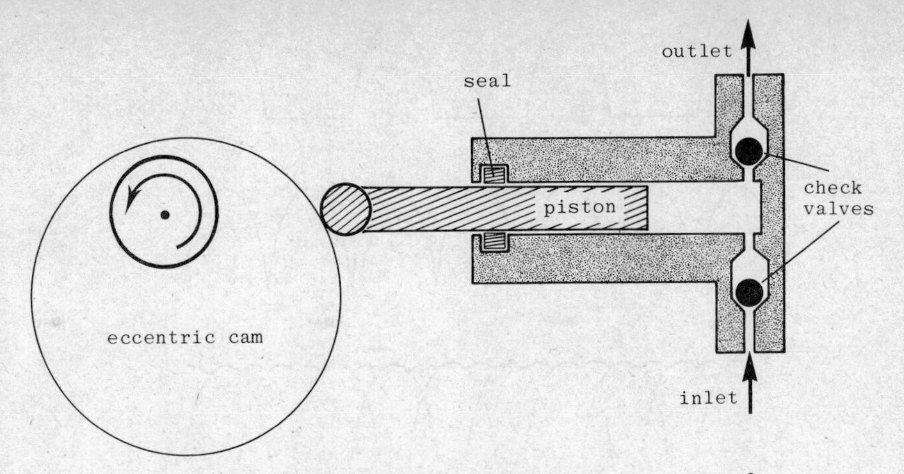

Figure 9.3. Reciprocating piston pump, displacement around 0.1 cm^3.

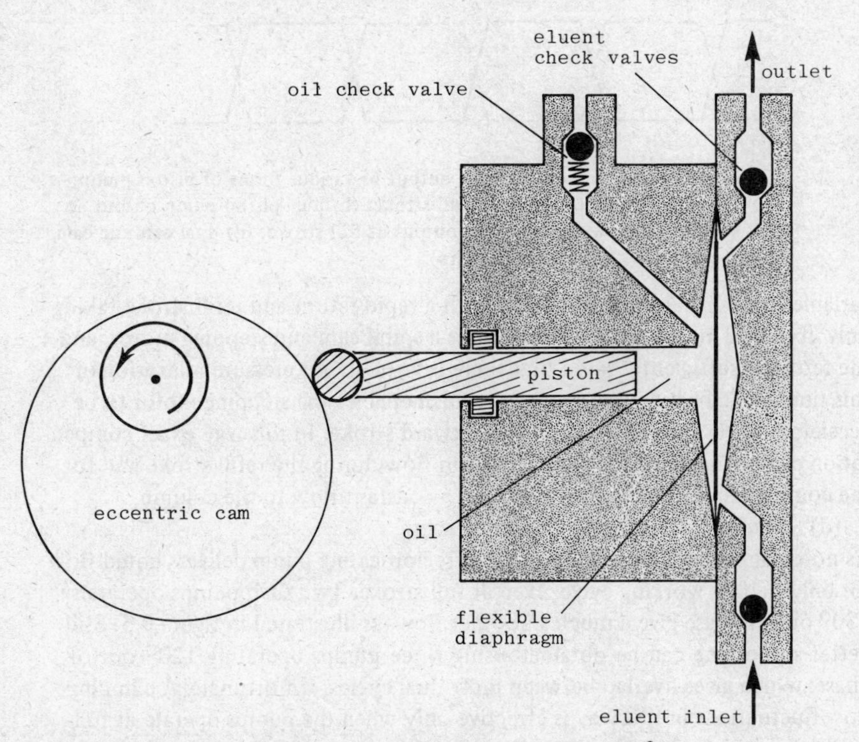

Figure 9.4. Diaphragm pump, displacement around 0.1 cm^3.

stroke. By using a sensitive pressure transducer an accurate measure of flow rate is obtained, and can be used with the feed-back system to pre-set the flow rate to any desired value. With two pumps controlled in this way gradient elution can be performed.

A recent development in the design of single-head pumps for HPLC uses a

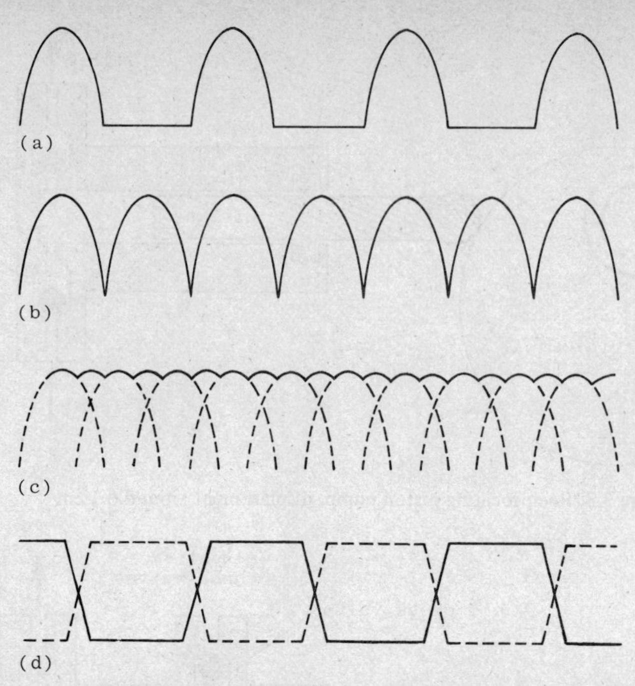

Figure 9.5. Variation with time of output of various forms of piston pump: (a) single-piston pump output at full stroke; (b) dual-piston pump output at full stroke; (c) triple-piston pump output at full stroke; (d) dual cardioid-cam pump output (with change-over ramp).

variable-speed linear drive combined with a rapid return and refill stroke taking only about 20 ms. This drive system uses a spiral cam and stepping motor, and the return is sufficiently rapid that there is virtually no pressure relaxation in this time. An adjustment is provided, which enables the stepping motor to be accelerated over the early part of the forward stroke. In this way exact compensation can be made for the slight hiatus in flow during the refill stroke and for the compression of the liquid, to ensure a constant flow to the column.

(d) *Multiple-Head Reciprocating Pumps*

As noted above, the circular-cam-driven reciprocating pump delivers liquid flow for only half its working cycle, even at full stroke. Two such pumps operated 180° out of phase give a much smoother flow, as illustrated in figure 9.5. Still better smoothing can be obtained using three pumps operating 120° out of phase, which gives overlap between individual cycles. Unfortunately, damping flow fluctuations in this way is effective only when the pumps operate at full stroke. If smooth flow is to be maintained over a range of flow rates then the flow rate can be changed only by altering the cycling time, but then pulsation may again become troublesome at low stroking rates.

The problem of designing small-volume reciprocating pumps that give completely uniform and adjustable output was solved by development of dual-head piston pumps driven by specially designed cams (figure 9.6). The modified cardioid-shaped cams induce linear displacement of the driven piston propor-

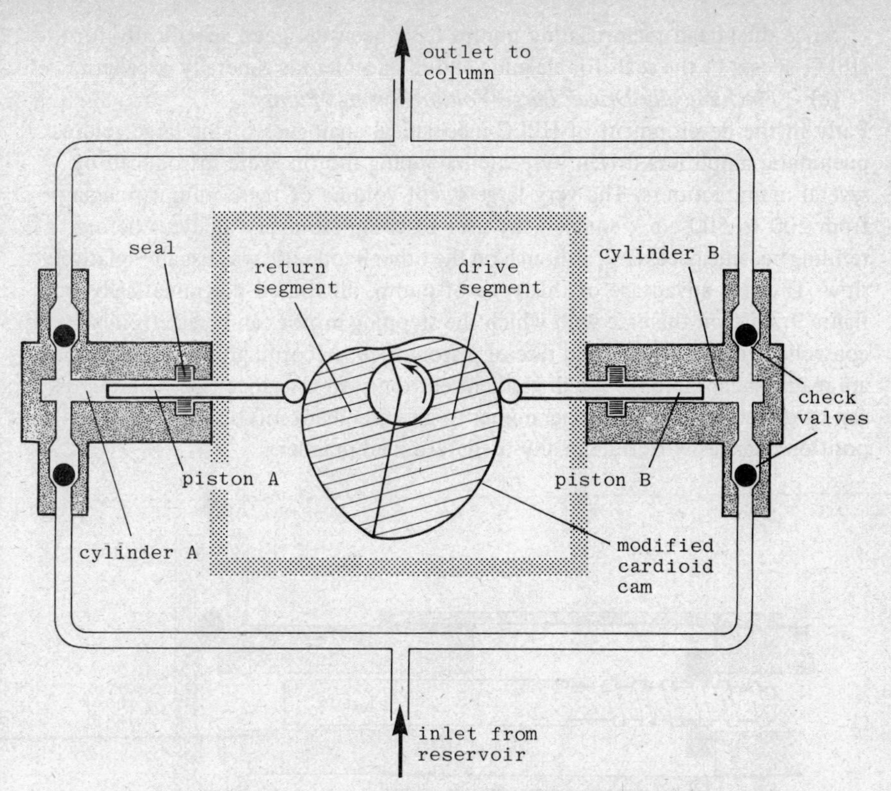

Figure 9.6. Dual-piston reciprocating pump driven by modified cardioid cam, displacement about 0.1 cm^3.

tional to the angle of rotation of the cam. Accordingly, twin heads driven by such cardioid cams and operated 180° out of phase would produce a completely pulse-free output flow if they were perfectly matched. In practice there will always be some mismatch, which will produce a slight pulse at the change-over points in the cycle. The effect of this is virtually eliminated by arranging a driving cycle of more than 180°, which allows a gradual take-over of one pump from the other (figure 9.5d). Several commercial versions of this class of pump are available, and they are probably the most widely used type at the present time.

Since dual-piston pumps are driven by stepping motors, their output is readily controlled electronically, and they can be programmed to give a flow varying with time. Two or more such pumps can therefore be used with an electronic programmer to produce composition gradients, and many modern gradient elution systems work on this principle. Alternatively, it is possible with most dual-piston pumps to generate changes in eluent composition in the low-pressure (inlet) side of the pump. This is because they have very low swept volumes and do not require down-stream pulse dampers. In principle this should be a much cheaper way of generating gradients than using two expensive pumps and an associated programmer.

Since dual-head reciprocating pumps have been designed specifically for HPLC, access to the seals for cleaning and replacement is generally excellent.

(e) *Mechanically Driven Large-Volume Syringe Pumps*

Early in the development of HPLC mechanical analogues of the large-volume pneumatic amplifiers driven by geared stepping motors were introduced by several manufacturers. The very large swept volume of these pumps, ranging from 200 to 500 cm^3, enabled the user to complete many analyses before refilling became necessary, although on the other hand refill was usually relatively slow. The real advantage of this type of pump, illustrated diagramatically in figure 9.7, lies in the ease with which the stepping motor can be electronically controlled to give any desired rate of piston drive. Accordingly, syringe pumps are readily adaptable for use in gradient elution systems, since two such pumps can be driven by a master programmer to give an eluent mixture whose composition changes with time in any predetermined manner.

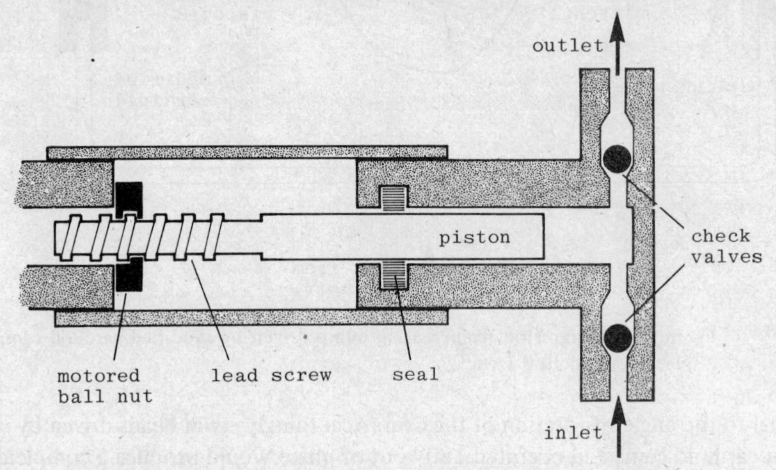

Figure 9.7. Screw-driven syringe pump, displacement 50 to 500 cm^3.

Most of the large syringe pumps do not allow ready access to the high-pressure seals for cleaning and replacement, although the wear rate of the seals is low, unlike those of reciprocating pump seals. A serious problem with this class of pump arises from the compressibility of liquids; this is discussed in the next section.

9.3 COMPRESSIBILITY EFFECTS

Much play has been made in both the scientific (1) and commercial literature of the relative advantages of the different types of pumping system that are supposed to produce a constant flow of eluent by using a piston driven at a closely controlled rate. Only belatedly have the consequences of liquid compressibility been appreciated in this context.

Most of the liquids used as eluents are reduced in volume by between 0.01 and 0.015% per bar; the only major exception is water, which has a compressibility of 0.004% per bar. At a column inlet pressure of 100 bar eluents are therefore compressed by from 0.5 to 1.5%, while at 200 bar they are compressed by from 1 to 3%. The effect of compressibility on the performance of large syringe pumps has been discussed at length by Guiochon (1) and Achener (2). A simple example will make the problem clear. Suppose one has a 200 cm^3 syringe pump that is intially full of eluent with a compressibility of 0.01% per bar, and suppose that a flow rate of 1 cm^3 min^{-1} is required with a column whose operating pressure at this flow rate is 100 bar. If the pump is not initially pressurised it is clear that 2 cm^3 of piston displacement will be required simply to bring the liquid to the desired pressure without any output flow. When output flow is allowed to the column then the build up of pressure and flow will be delayed. Indeed it is not difficult to show that, for a nominal flow rate of 1 cm^3 min^{-1}, the actual flow at the column inlet will take 4.6 min to reach 90% of its final value and 9.2 min to reach 99% of its final value. Clearly, over this not insignificant period the pump is not delivering the intended flow rate, but something less. This difficulty is overcome with a single pump by pre-pressurising the eluent inside the pump chamber against a closed downstream valve. If the valve is opened when sample elution is started there will be no further delay in reaching the set flow rate, provided this is the same as before and the downstream resistance has not changed, but this operating pressure must first be found in an independent experiment.

When using two such pumps to generate a gradient the problem is more complex, because in all probability the viscosity of the eluent will change during elution and the pressure drop across the column will also change. In this case compressibility effects can be mitigated by placing a high-pressure restriction valve after the mixing chamber as shown by Abbott (3), as long as the valve back-pressure is several times that of the packed column.

One might imagine that these effects would be absent from cam-driven dual-head reciprocating pumps, but here the problem is different. Suppose that an analysis is proceeding at a flow rate of 1 cm^3 min^{-1} and a column inlet pressure of 100 bar. During the linear part of the forward stroke of piston A cylinder B is refilled at ambient pressure; when piston B starts on its forward stroke it will have to displace 1% of its swept volume before the liquid in cylinder B is sufficiently pressurised to open the output non-return valve B and hence pass into the delivery line leading to the column. During this part of the stroke there will be no flow from pump B, and therefore a small but significant change in flow rate will occur at each change-over. This is probably chromatographically unimportant in isocratic elution at moderate pressures, but in gradient elution, especially where one of the eluent components is present as a very small proportion of the mixture, the change in flow rate over these change-over periods could have a noticeable effect on the actual composition of the eluent mixture.

These considerations may be thought fairly academic, and indeed they probably are as long as HPLC systems are run at column inlet pressures below 100 bar. However, as soon as the very high pressure capability of modern pumps is put to use, compressibility effects will immediately become much more pronounced.

Recognition of the compressibility problem by manufacturers of dual reciprocating pumps has led them to develop flow-feedback control systems. These continuously monitor the pressure drop across a restriction in the eluent delivery line and electronically control the stepping rate of the driving motors to keep that pressure constant.

In conclusion, there appears to be no clearly superior pumping system in the current HPLC spectrum, and there is only one, rather expensive system that can guarantee that the dialled flow is genuinely passed into the head of the column. In the other cases, the flow control setting really determines only the rate of volume displacement by the piston. For a number of reasons this may well not be the same as the actual flow of liquid to the column. Nevertheless current pumps are in general satisfactory at the level of precision expected for current HPLC. The least expensive pumps are quite adequate for routine analyses, and there is no overriding reason to prefer either constant pressure or constant flow operation. More expensive dial-a-flow pumps, with or without feedback control, do offer greater flexibility and convenience but do not provide any great increase in precision.

9.4 GRADIENT ELUTION SYSTEMS

Gradient elution, otherwise termed solvent programming, is used when the range of retention of the components of a mixture is so large that one cannot conveniently elute them all with a single eluent. It is then necessary either to use several eluents sequentially or to change the composition of the eluent in a pre-determined way. In normal-phase chromatography (e.g. adsorption chromatography or partition chromatography with a polar stationary phase) the eluent is made progressively more polar, whereas in reversed-phase chromatography (e.g. chromatography on a bonded support with an aqueous or alcoholic eluent) the eluent is made progressively less polar. Gradient elution is used in situations similar to those demanding temperature programming in gas chromatography. The majority of commercial HPLC systems generate these eluent composition changes by programmed mixing of two liquids.

The overriding consideration for any satisfactory gradient-producing system is that it operates in a reproducible fashion. In practice this means that the volume ratio of the two components in the eluent must be controlled to within 1 or 2% at any stage in the programme. Whilst this degree of control is relatively easy to achieve when the different components are mixed in roughly equal proportions, it becomes progressively more difficult as the ratio becomes more extreme. There are few commercial gradient systems that operate really reproducibly at composition ratios of less than 1:20 and more than 20:1. Failure to control the mixture composition to a sufficient degree of precision will result in irreproducible retention times as well as variations in the resolution of mixture components. In addition to being able to change the eluent composition reproducibly during the chromatographic analysis, a gradient system must also have provision for reconditioning the column by bringing the eluent composition back to its original value in readiness for the next run.

Although programming the flows from two electronically controlled mechanical pumps (say, of the types described in sections 9.2d and e) is the simplest way to generate a solvent gradient, it is a fairly expensive system and by no means the only possible method. Classical amino-acid analysers use either low-pressure valves to control the selection of a sequence of eluents in separate reservoirs, or a sequence of mixing vessels. A wide variety of solvent programmes can be generated by adjusting the timing of the valves or of the number and volumes of the mixing vessels. A very simple system of the first type, which has been used in HPLC, consists of a dozen or so reservoirs filled with a range of eluent liquids that are tapped sequentially (4). The major limitation of systems based upon multiple reservoirs is that setting up is tedious and, in some cases, may have to be repeated for each new analysis. A simple system of the second type used a single fully stirred mixing chamber, which produces an exponentially changing solvent composition; this can be used either on the high- or low-pressure side of the pump. After each run the chamber is refilled with the initial solvent. Low-pressure gradient systems should not be neglected, since they can be used with modern dual-head pumps of low swept volume and can be constructed fairly cheaply. They have been discussed in detail by Litenau and Gocan (5).

A commercial system, which enables any programme of solvent change to be generated, is illustrated in figure 9.8. A single pump containing the first eluent A is used in conjunction with a 200 cm^3 holding coil containing the second eluent B. With the solenoid-controlled valves (a), (b) open and (c), (d), (e) closed, the coil may be washed out and filled from the bottom with liquid B. In gradient operation (a), (b) are closed and (c) is open. Valves (d), (e) then cycle alternately under the control of an electronic programmer. When (d) is open and (e) closed, liquid A flows to the column, whereas when (d) is closed and (e) open liquid A is directed into the coil and displaces B, which flows to the column. With a cycle time of a few seconds and a suitable mixing chamber any mixture from about 5 to 95% of either liquid can be produced with good precision at currently used flow rates. The system can be programmed by any suitable electronic system that can accurately control the valve cycles.

Some general comments on gradient systems and their choice may be made. The commercial literature frequently praises the ease with which particular equipment can be used to set up and deliver complicated solvent composition programmes, but the fact remains that designing an optimum gradient is very time-consuming, and it is seldom worthwhile to invest the time and energy required for a single or even a few separations. It is only for widely used routine analyses, such as amino-acid analysis, that optimisation of a gradient is profitable. Nevertheless the ability to dial a mixture of any desired composition for a particular analysis or to run a simple gradient for "scouting" purposes may be of great value and may add substantially to the usefulness of a chromatograph. In particular, it can greatly accelerate the search for the best eluent composition for isocratic separation of a particular mixture.

Choice of a commercial gradient generating system is dictated to a great extent by the type of pump employed. Thus, manufacturers of digitally controlled syringe and dual-head pumps usually offer a twin-pump system controlled by an electronic programmer, but such systems are inevitably expensive because

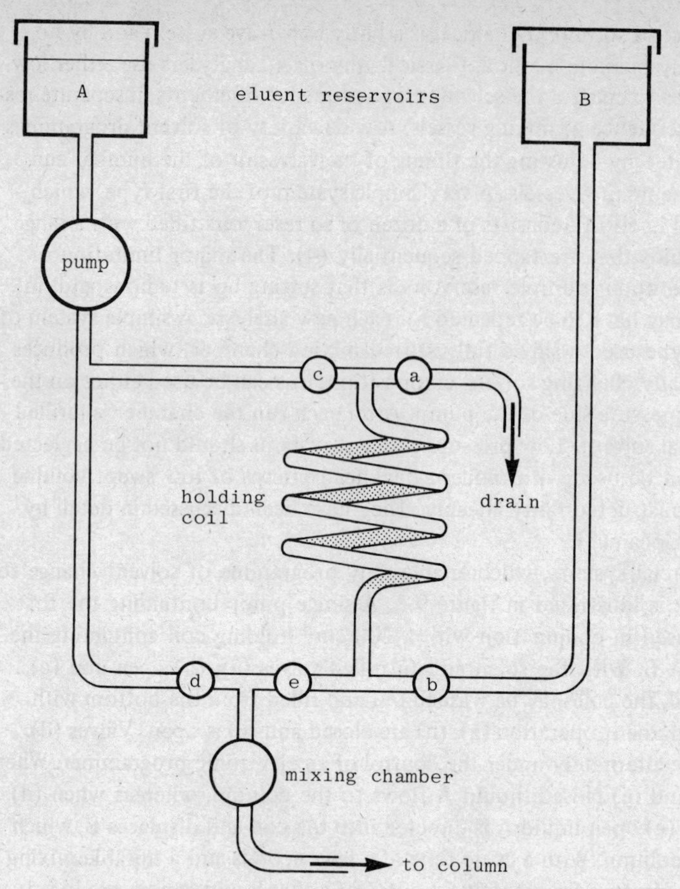

Figure 9.8. Gradient-forming system employing a single pump and holding coil.

the pumps are themselves expensive. Alternatively, pumps having a low swept volume, and in particular dual-head reciprocating pumps, may be used with one of the many gradient generators feeding the inlet. Commercial chromatographs fitted with pneumatic intensifier pumps are offered with variants of the holding-coil system, which should be fairly inexpensive in principle although regrettably this is not generally true in practice.

REFERENCES

1. Martin, M., Blu, G., Eon, C. and Guiochon, G., *J. Chromatogr. 112* (1975) 399.
2. Achener, P., Abbott, S.R. and Stevenson, R.L., *J. Chromatogr. 130* (1977) 29.
3. Abbott, S.R., Berg, J.R., Achener, P. and Stevenson, R.L., *J. Chromatogr. 126* (1975) 421.
4. Scott, R.P.W. and Kucera, P., *J. Chromatogr. Sci. 11* (1973) 83.
5. Litenau, C. and Gocán, S., *Gradient Liquid Chromatography.* John Wiley, London and New York, 1974.

10. DETECTORS AND REACTOR SYSTEMS

The function of the detector in any chromatograph is to monitor the concentration or quantity of the sample components emerging from the column. It should provide an electrical signal proportional to concentration or quantity, which can be used to operate a data recording or processing system such as a potentiometric recorder, an electronic integrator or a computer system. It must perform these functions with high precision, high sensitivity and high stability, and its operation must not significantly attenuate the separation achieved by the column.

10.1 CLASSIFICATION OF DETECTORS

Three main types of detector are currently used in HPLC: (a) selective property detectors such as the UV photometer; (b) bulk property detectors such as the refractive index monitor; and (c) solute/eluent separation systems such as the wire transport-FID detector. Because of the ease with which high-performance analytical columns can be overloaded, satisfactory detectors must be capable of detecting less than 1 part of solute in 10^6 parts of eluent. Unfortunately, because of the temperature sensitivity of bulk properties of liquids, and the rather small differences between their values for solutes and eluents, most bulk property detectors only just meet this requirement. Selective detectors, on the other hand, are from 100 to 100,000 times as sensitive to favourable solutes as to eluents, and are undoubtedly to be preferred whenever they are applicable.

The sensitivity requirement of an HPLC detector must be combined with the need to keep the volume of the cell in which measurement is made to a minimum, and to design the cell so that it is well swept by the flow of eluent. The standard 10 mm long, 1 mm bore (8 mm^3) flow-cells used in photometric detectors are in fact only just adequate for use with short (100 mm), wide (5mm bore) columns packed with 5 μm particles, as shown in chapter 8.

10.2 CRITERIA FOR CHOICE OF DETECTOR

In order to choose a detector for any given application one must be able to compare detectors on the basis of a uniform set of criteria. Such criteria have been discussed in detail by Scott (1), the more important detector parameters being the following:

(a) The noise level of the detector, N_D. This is given in terms of the property measured by the detector.

(b) The noise equivalent concentration, C_N. This is the concentration of solute that produces a signal equal to the noise level. Since this is a solute-dependent property, the solute used to define C_N must be specified and the value of its characteristic property given.

(c) The dynamic range, R_D. This is the ratio of the maximum to the minimum concentration of solute (i.e. the range of solute concentration) over which the detected property can be measured.

(d) The useful linear range of the detector, R_L. This is determined by plotting the log of the detector response against the log of the concentration of a typical solute. The gradient of such a plot is called the response index, r, and would be exactly unity for a perfectly linear detector. The gradient obtained in practice will generally be close to unity at low concentrations but will inevitably deviate from the ideal value at some finite concentration. The useful linear range of the detector, R_L, is defined as that range of solute concentration over which the response index lies between 0.98 and 1.02.

(e) The volume, V_{cell}, in which the detected property is measured.

(f) The length and bore of any tubing necessarily associated with the detector, e.g. that required for the heat exchanger in a refractive index monitor.

(g) The band dispersion, w_{det}, produced by the detector cell, measured in volume units (see sections 7.1 and 10.8, below).

(h) The band dispersion, w_{tube}, produced by any heat exchanger or other tubing necessarily associated with the detector, e.g. the reaction coil or packed bed of a post-column reactor (see section 10.7, below).

In the following sections these parameters will be discussed for different detectors. Table 10.1 summarises the information on these parameters as far as it is available.

Table 10.1. Detector characteristics.

Type of detector	N_D	C_N [a] (g cm^{-3})	R_L [b] (orders of magnitude)	V_{cell} (mm^3)
UV photometer	10^{-5} AU	10^{-10}	5	8
Spectrophotometer	10^{-4} AU	10^{-9}	4	
Fluorimeter	equivalent to 10^{-7} AU	10^{-12}	5	8–16
Coulometric	10^{-10} amp	10^{-10}	5	30
Polarograph	10^{-10} amp	10^{-8}	4	1–5
RI monitor	10^{-8} RI units	10^{-7}	4	20–50[c]
Transport		10^{-6}	3	
Conductivity meter		10^{-6}–10^{-8}	3	1–10
Radioactivity meter	5 counts per min		large	200
Dielectric constant		10^{-6}	3	1–10
Nephelometric		10^{-6}	2	20

[a] In favourable cases for selective detectors.

[b] The dynamic range R_D is generally about 10 times the linear range R_L.

[c] Including heat exchanger.

10.3 SPECIFIC PROPERTY DETECTORS

(a) *Flow Photometers*

The flow photometer has been a feature of HPLC instrumentation since its inception, and remains the most popular of the HPLC detectors. The great majority monitor the attenuation of a beam of UV light when it passes through a flow-cell with an optical path of 10 mm length and 1 mm bore, the cell being capped with silica or sapphire windows.

When monochromatic radiation is used the absorbance of light obeys Beer's Law. The absorbance, A, is defined by

$$A \equiv \log_{10}\frac{I_0}{I} = \epsilon c l \qquad (10.1)$$

where I_0 is the intensity of light that would emerge from the cell if it contained a completely transparent liquid, and I is the intensity of light emerging under the same conditions when an absorbing sample is present; ϵ is the decadic molar extinction coefficient in units of $mol^{-1}\ dm^3\ cm^{-1}$, c the solute concentration in $mol\ dm^{-3}$ and l the optical path length in cm.

Very often the optical absorbing power of a substance in solution is quoted not in terms of its molar extinction coefficient, ϵ, but in terms of its $A^{1\%}_{1\ cm}$ value. This is the absorbance, A, of a 1% w/v solution (1 gm in 100 cm^3) of the solute when measured with an optical path length of 1 cm. $A^{1\%}_{1\ cm}$ and ϵ are related by

$$A^{1\%}_{1\,cm} = \frac{10\epsilon}{MW} \qquad (10.2)$$

Thus for a solute of molecular weight 100 and $\epsilon = 10^4\ mol^{-1}\ dm^3\ cm^{-1}$, $A^{1\%}_{1\ cm} = 10^3$.

The first commercial flow photometers all used low-pressure mercury arc sources, since such lamps have a long life (around 5000 hours) and give a highly stable and intense output of the resonance radiation at 254 nm wavelength. This line accounts for about 90% of the emission and only the simplest optical filter is required to isolate it. The radiation falls conveniently in a region of relatively high absorption of many organic molecules (aromatics, conjugated ketones, heterocyclic aromatics, etc.) and, accordingly, relatively simple photometers continue to be widely used in HPLC. The noise level, N_D, of the best double-beam instruments is about 10^{-5} absorbance units (AU), which corresponds to a noise equivalent concentration, C_N, of $10^{-10}\ g\ cm^{-3}$ for a solute of molecular weight 100 and molar extinction coefficient $\epsilon = 10^4\ mol^{-1}\ dm^3\ cm^{-1}$. The linear range of absorbance measurement is roughly 10^5, that is from the noise level $A = 10^{-5}$ to $A = 1$. For absorbance above unity it becomes increasingly difficult to eliminate extraneous effects such as stray light and electrical background signals.

In the quest for increased sensitivity and specificity the first improvement on the simple fixed-wavelength photometer was to replace the mercury resonance lamp by a more intense deuterium lamp giving a continuous output from 180–400 nm (or a tungsten filament lamp with an output from 400–700 nm). These lamps were used in combination with a manually adjustable grating, which allowed selection of the detection wavelength to match the absorption maximum of the sample. Since the light intensity transmitted is proportional to the band width of the radiation a compromise is always required, for with too wide a band Beer's Law is not obeyed, while with too narrow a band the noise level is increased. The normal compromise is to use a band width of 2–5 nm. In general, deuterium lamps have a sufficiently stable output that the best single-beam instruments have noise levels only marginally higher than those of the fixed-wavelength photometers. Their price is also only slightly higher.

While single-wavelength monitoring can tell the chromatographer the elution order and concentration of eluted components, it cannot identify them. With one of the classical identification tools of chemistry being spectrophotometry, an obvious development was that of the automatic scanning spectrophotometric detector. Since scanning normally requires a considerable time, the flow of eluent must be arrested while the component of interest is held in the cell and the photometer scan carried out. For this purpose a double-beam photometer is required, along with facilities for stopping the flow, initiating the scan over the required range, resetting the wavelength to that required for detection, and finally restarting the flow. It is also convenient to use two recorders, one to obtain the chromatogram and the other to record the spectra of the components of interest. Such automatic scanning systems are inevitably expensive, costing two to four times as much as simple spectrophotometric instruments. The typical band width for detection is 2 nm, while that for identification should be adjustable.

The latest development (2) in the field of photometric detection is that of the diode array spectrophotometer, in which a beam of "white" radiation is passed through the flow-cell (rather than a beam of monochromatic radiation). The light emerging from the flow-cell is diffracted by a grating so that it falls upon an array of photodiodes, each diode receiving a different wavelength. In this way an individual signal can be obtained for every 2–5 nm of the wavelength range. One complete spectrum can be stored every two seconds, and thus several spectra can be stored for even the most rapidly eluting peaks in HPLC. The diode array spectrophotometer can be used for quantitative analysis even when solute peaks are only partially resolved. Their sensitivity and noise level is only slightly inferior to that of more conventional spectrophotometers. At the time of writing they are not commercially available, and will presumably be expensive when they do emerge.

Another interesting photometric detector is the infra-red photometer. Such instruments are of intermediate sensitivity (between that of a refractometer and a UV photometer). They are marketed by one or two manufacturers, and have considerable potential in exclusion chromatography and some other specialised applications where selectivity is a prime consideration.

The typical cell used in HPLC flow photometers has a length of 10 mm and a bore of 1 mm. These dimensions, as noted in chapter 8, result in an additional peak dispersion of between 30 and 40 mm^3. Such measuring cells are only just good enough to be used with short (100–120 mm), wide (4–6 mm bore) columns packed with small ($d_p \approx 5\ \mu m$) particles. If any significant improvements in column technology occur to reduce peak volumes below the 60–100 mm^3 achieved in the best current HPLC practice, it will become necessary to decrease all the dimensions of the measuring cell. Any decrease in path length will reduce the sensitivity, while a decrease in the bore, which would be the most effective way of reducing the additional dispersion, will cut down the transmitted light intensity and so increase the noise level. Any such changes in cell design will therefore require appropriate changes in the basic photometer design.

There are two variants of the classical Z-pattern flow cell, which is illustrated in figure 10.1. The design shown in figure 10.2 was developed by Kirkland (3).

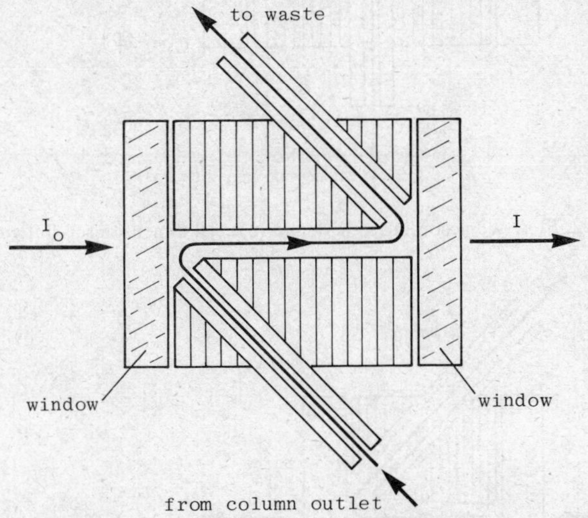

Figure 10.1. "Classical" Z-pattern photometer flow-cell; optical path length = 10 mm; bore = 1 mm; V_{cell} = 8 mm^3.

The split-flow was introduced in the belief that it would reduce disturbances arising from the eluent flow, which can produce noise and drift, but it is somewhat doubtful if there is much difference in practice between the flow sensitivity of the two types. Another attempt to remove flow-derived disturbances is to use a tapered cell (4) as shown in figure 10.3.

It is now known that flow-derived disturbances in flow cells result almost entirely from changes in the refractive index (RI) of the eluate. Because of the non-uniform flow across the section of a circular hole, any time-dependent change in the RI of eluate will be reflected in a variation of RI across the channel of the photometer cell. This change distorts the light beam, and may result either in an increase or decrease of the light falling on the photocell monitoring the beam. The most common manifestation of this effect is a differential peak, which is observed when a non-absorbing solute passes through the cell. This is often noted around the time expected for elution of an unretained solute, especially when the sample has been dissolved in a solvent different from the eluent. Particularly good examples of this effect are shown in figures 3.8 and 5.9.

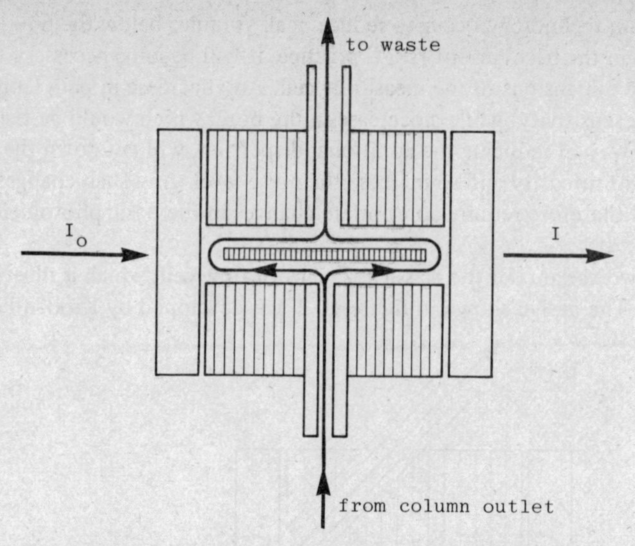

Figure 10.2. Flow-cell according to Kirkland (3): cell dimensions as in figure 10.1.

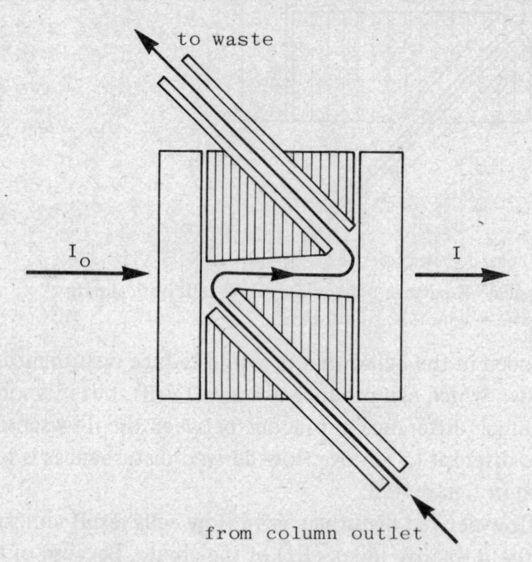

Figure 10.3. Taper bore flow-cell according to Little and Fallick (4): optical path length = 6 mm; bore at entry = 1 mm; bore at outlet = 1.5 mm; $V_{cell} = 8\ mm^3$.

The role of turbulence in flow cells is also important. With a transparent mock-up of a photometer cell the flow pattern at typical eluent flow rates of around 1 $cm^3\ min^{-1}$ can be readily seen with coloured samples. As would be expected with the classical Z-pattern cell, there is turbulent flow at the point of fluid entry, but this rapidly decays into laminar flow as the cell length is traversed, as shown in figure 10.4. Laminar flow ensures that when a sample of different refractive index enters the cell, the new solution appears first in the centre of the cell and last at the cell wall. This creates a refractive index

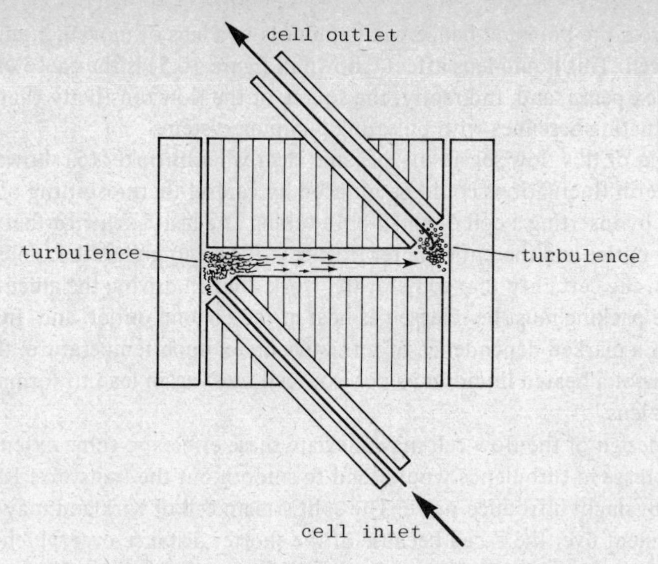

Figure 10.4. "Classical" flow-cell flow patterns showing turbulence at entry from 0.25 mm bore tubing connection decaying to laminar flow in centre, and turbulence forming at outlet connection.

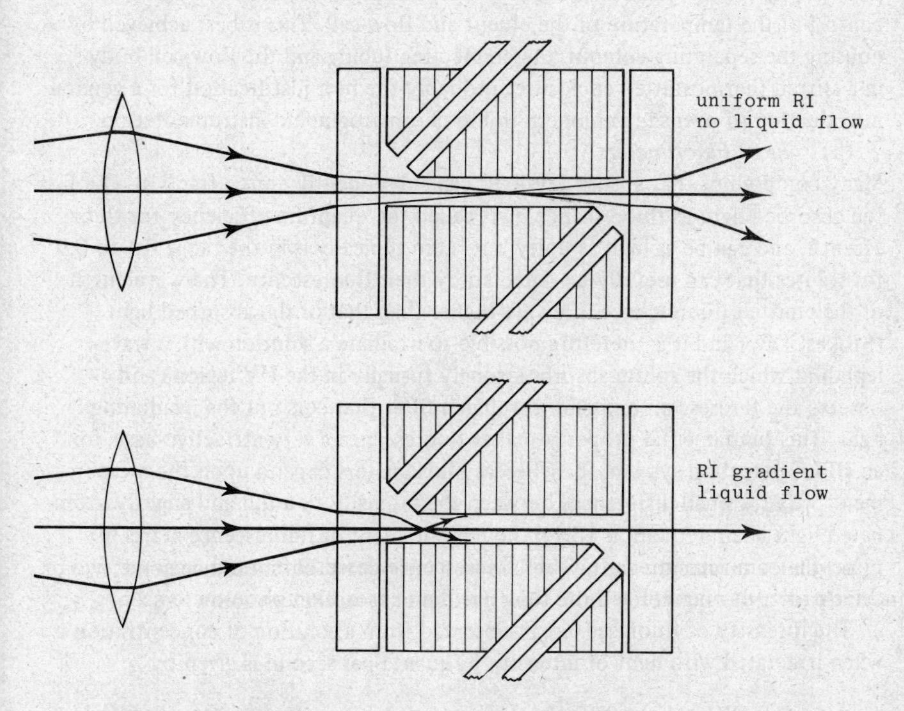

Figure 10.5. "Classical" flow-cell showing effect of refractive index gradient (smallest at centre of bore, largest at wall of bore) as produced by laminar flow/temperature gradient.

gradient across the bore and hence what amounts to a lens of moving liquid within the cell. This liquid-lens effect (shown in figure 10.5) is the cause of refractive index peaks, and, indirectly, the source of the flow sensitivity that gives rise to fluctuating baselines with pulsating pumping systems.

The origin of this flow sensitivity became clear when Brooker (6) showed that short-term fluctuations could be reduced by careful thermostatting of the detector or by inserting a coil of small-bore tubing as a heat exchanger between the column outlet and the photometer flow-cell. If one combines these observations with the certainty that some of the work done in driving the eluent through the packing must be released as heat at the column outlet, and, further, that there is a marked dependence of refractive index upon temperature, then laminar flow of a heated liquid in a cool flow-cell must again lead to formation of a "liquid lens".

Careful design of the flow-cell may alleviate these effects to some extent. Thus an increase in turbulence would tend to smooth out the transverse RI gradients but might introduce noise. The split stream cell of Kirkland may be an improvement over the Z-cell because of the shorter distance over which laminar flow can develop. The tapered cell may be an improvement because the wall regions where flow is slowest are not in fact "seen" by the photocell of the detector. While some further improvements in flow-cell design are no doubt possible, by far the most effective way of reducing flow effects is by careful control of the temperature of the eluent and flow-cell. This is best achieved by housing the separating column, the connecting tubing and the flow-cell body in a stirred thermostatted enclosure: probably the best justification for a general introduction of ovens into modern liquid chromatographic instrumentation.

(b) *Flow Fluorimeters*

Many compounds that absorb UV light can subsequently emit a fraction, Q, of the absorbed light as fluorescence. Q is called the quantum efficiency for fluorescence, and can be as high as unity but more typically is in the range 0.1 to 0.9 for solutes that can usefully be detected by their fluorescence. The wavelength of the emitted fluorescence is always higher than that of the absorbed light (Stokes' Law) and it is therefore possible to irradiate a solution with a wavelength at which the solute absorbs strongly (usually in the UV region) and observe the fluorescent emission through a filter that cuts out the irradiating light. This fundamental property makes fluorescence a very attractive basis for an HPLC detection system, for whereas photometers depend upon the measurement of fairly small differences between the intensity of a full and slightly attenuated light beam (equation 10.1), the measurement of fluorescence starts in principle from zero intensity. Thus fluorescence detectors should be some two or three orders of magnitude more sensitive than absorption photometers.

The intensity of fluorescence, F, observed from a solution of concentration c when irradiated with light of intensity I_o quanta per second is given by

$$F = QKI_o\,(1 - 10^{-\epsilon cl}) \qquad (10.3)$$

where K is the efficiency of collection of the fluorescence, ϵ is the molar extinction coefficient of the solute and l the path length. This expression produces a concave dependence of F upon c. However, at low enough values of c the dependence is linear and is given by

$$F = 2.303\ QKI_o\ \epsilon cl \qquad (10.4)$$

Under these conditions the response index, r, is exactly unity, but falls to 0.98 when the absorbance is about 0.02 and to 0.95 when the absorbance is about 0.05.

Flow fluorimeters have been available since the earliest days of commercial HPLC, but only recently have the problems of flow-cell design been adequately resolved. The newer detectors do indeed have very much higher sensitivies than photometers, and for favourable samples give noise equivalent concentrations, C_N, as low as 10^{-12} g cm^{-3}. They are particularly useful when used in conjunction with sample derivatisation: thus, for example, amines can be detected with high sensitivity and specificity as their fluorescent dansyl derivatives (dansyl = dimethylaminonaphthalenesulphonyl), or by post-column reaction with o-phthalaldehyde (see figure 6.4). The increase in sensitivity when using fluorescence is, however, gained at some cost because of the constraints imposed by flow-cell design and construction and by the very low levels of light emitted. Figure 10.6 shows one design of fluorimeter cell with a light absorption path of 20 mm by 1 mm bore (V_{cell} = 16 mm^3). The fluorescence emerges randomly from the irradiated sample. It is collected by a parabolic mirror, which focusses the light onto a photomultiplier cell after passage through a filter that cuts out the exciting radiation. With this geometry the absorption of the irradiating

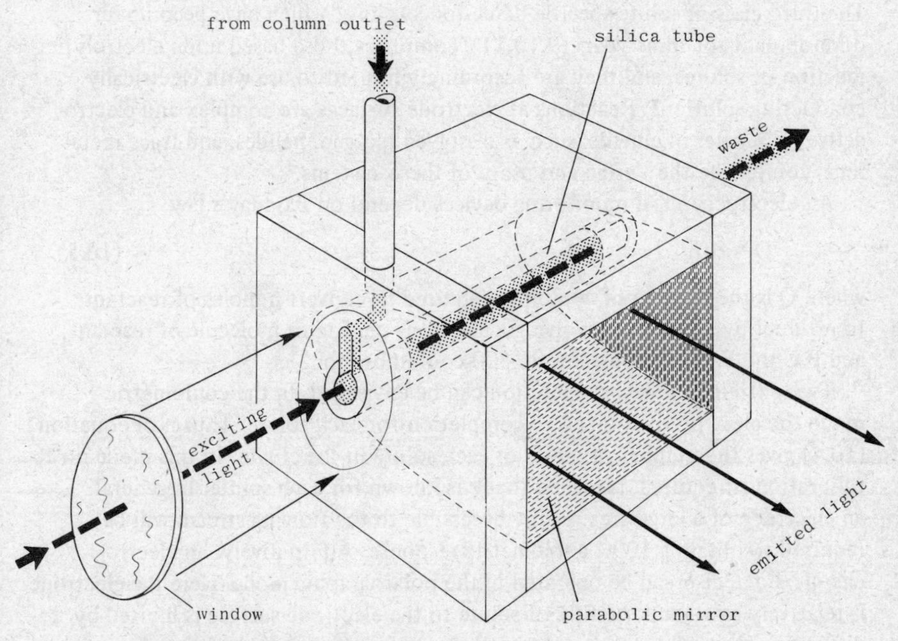

Figure 10.6. Fluorimeter flow-cell: optical path for exciting light = 16 mm; bore of tube = 1.2 mm; V_{cell} = 20 mm^3.

light can be measured independently of the intensity of fluorescence. Thus dual fluorescence/absorption detection is possible, and a number of commercial instruments have this dual facility.

The principle of a multichannel optical analyser using a diode array, as recently used for photometry (2), has also been applied to fluorimetry by Saner *et al.* (7). The same advantages apply, since fluorescence spectra are equally characteristic of the sample molecule and since qualitative as well as quantitative information is presented in the data.

The principal limitation of conventional flow-fluorimetric monitors is low emitted light levels. Since the number of light quanta emitted is proportional to the absorbed light energy (equation 10.2) an increase in the intensity of the exciting radiation will increase the intensity of the fluorescence proportionally. Two methods are generally available for substantially increasing light flux, the use of either a high intensity gas discharge lamp (e.g. xenon flash lamp) or a laser. Unfortunately, most good fluorophores require UV or near-UV exciting light, and UV lasers are not readily available; however, emission can be stimulated by simultaneous absorption of two photons of lower individual energy. In an interesting application of this principle Young (8) used two-photon excitation by a 4 W, 514.5 nm argon laser to detect some oxadiazoles. Two photons of this wavelength provide about the same activating effect as one photon of the 254 nm mercury lamp resonance line used in more conventional fluorimeters. Commercial development of this idea awaits a broad spectrum tunable laser, and hence will again be very expensive.

(c) *Electrochemical Detectors*

The third class of solute-specific detection systems, which have been under development for some years (9,10,11), comprises those based upon electrolytic reaction of solutes, and they are accordingly limited to use with electrically conducting solutions. Reactions at electrode surfaces are complex and electroactive impurities in eluents, such as dissolved oxygen, halides, and trace metal ions, complicate the design and usage of these systems.

All electrochemical monitoring devices depend on Faraday's law

$$Q = znF \qquad (10.5)$$

where Q is the number of coulombs required to convert n moles of reactant to product by a reaction involving z electrons per ion or molecule of reactant, and F is the Faraday constant, 96,500 coulombs mol^{-1}.

Two extreme modes of operation can be envisaged. In the coulometric mode the electrolysis is taken to completion for each solute. Thus n of equation (10.4) gives the number of moles of each solute in the eluate. In this mode no calibration is required, provided that z is known for each solute. In general an electrode of a large area readily accessible to the flowing stream will be required to obtain a 100% coulometric response. Alternatively, an electrochemical detector can be operated in the polarographic mode. Here the electrode is relatively small and the flux of solute to the electrode surface is limited by diffusion. The diffusion rate across the boundary layer to the electrode surface is proportional to the concentration of the solute in the bulk liquid. In the polarographic mode the electrode current is then proportional to concentration and the diffusion coefficient of the solute being oxidised or reduced, and calibration is therefore necessary. The coulometric efficiency of such devices is typically in the range 0.1 to 1%.

Both types of detector have been used successfully and a number of detectors are of intermediate coulometric efficiency.

(i) *Coulometric Detectors.* The variable measured in any electrochemical detector is, of course, electrical current; for a coulometric detector this will be given by

$$i = \frac{dQ}{dt} = zF\,\frac{dn}{dt} \qquad (10.6)$$

where dn/dt is the number of moles of solute passing into the device per second. Since a noise level of 10^{-10} amps can be achieved (12), the rate of passage of solute equivalent to the noise is about 10^{-15} mol s^{-1}. Assuming a flow rate of 10 mm^3 s^{-1} (0.6 cm^3 min^{-1}) a molecular weight of 100 and z = 1 gives $C_N = 10^{-11}$ g cm^{-3}, some 10 times better than the sensitivity of the best flow photometers.

Since the response from a coulometric detector is a function of the number of molecules flowing over the electroactive surface, the measured current will vary in direct proportion to the volumetric flow rate. A disadvantage of such a system is that it will be sensitive to minor perturbations, such as eluent flow due to pump fluctuations, etc., although the integral over any peak will not be affected. A more serious disadvantage of this system arises from the need to convert all the solute to product within the measuring cell. Since the rate of reaction at the electrode surface may be limited by the rate of diffusion of reactants through the solvent in the flow cell, great attention to the design is required.

In the flat plate cell design of Lankelma and Poppe (12), shown in figure 10.7, the diffusion problem is overcome by arranging that the electrode spacing is extremely small (around 50 μm). Then by using a large enough area (typically 80 X 7 mm) the residence time of any molecule of solute in the cell can be made much greater than the diffusion time across this layer. Thus for the cell just specified V_{cell} = 30 mm^3, while the time for diffusion is about one second for an aqueous solution. With typical flow rates of about 10 mm^3 s^{-1} (0.6 cm^3 min^{-1}) the diffusion time is about one-third of the residence time. This is adequate to ensure virtually 100% reaction of any electroactive species.

An alternative to the flat plate electrode system is a packed bed, where once again complete mass transfer to the electrode surface can be assured. The design of figure 10.8 was described by Buchta and Papa (13), but in use proved to give no better signal-to-noise ratio than a much simpler wire electrode polarographic detector. In theory, a packed-bed measuring cell should give a very large electroactive surface area in a configuration with minimal chromatographic zone broadening.

(ii) *Polarographic Detectors.* Despite the analytical advantages of coulometric

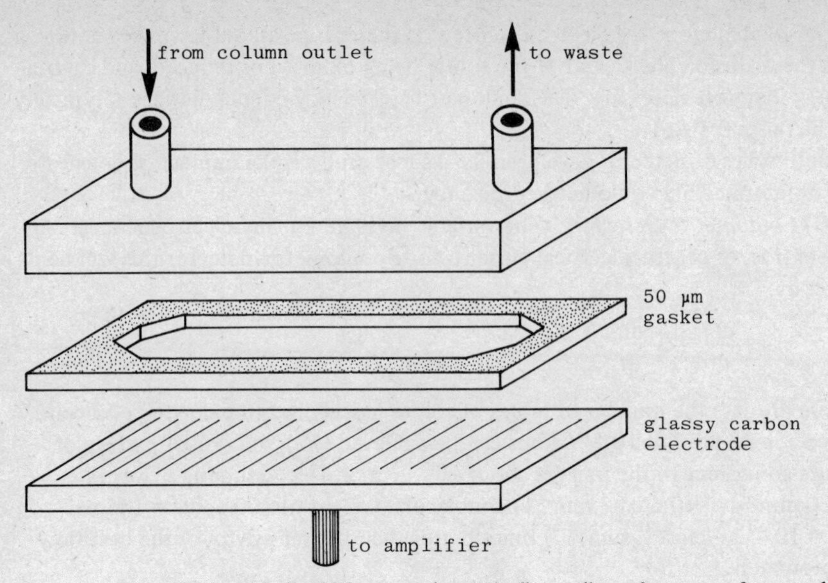

Figure 10.7. Thin liquid layer coulometric flow-cell: surface area of exposed glassy carbon electrode = 560 mm^2; aperture of gasket = 80 mm x 7 mm x 0.05 mm; V_{cell} = 28 mm^3.

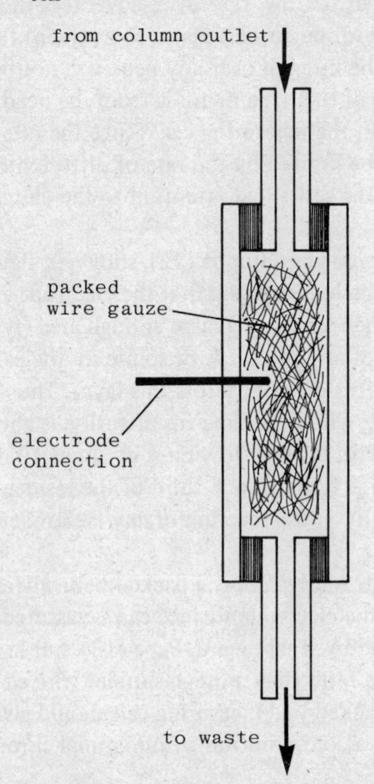

Figure 10.8. Packed-bed coulometric flow-cell: platinum wire gauze in 6 mm bore x 50 mm long tube; free volume = 560 mm^3.

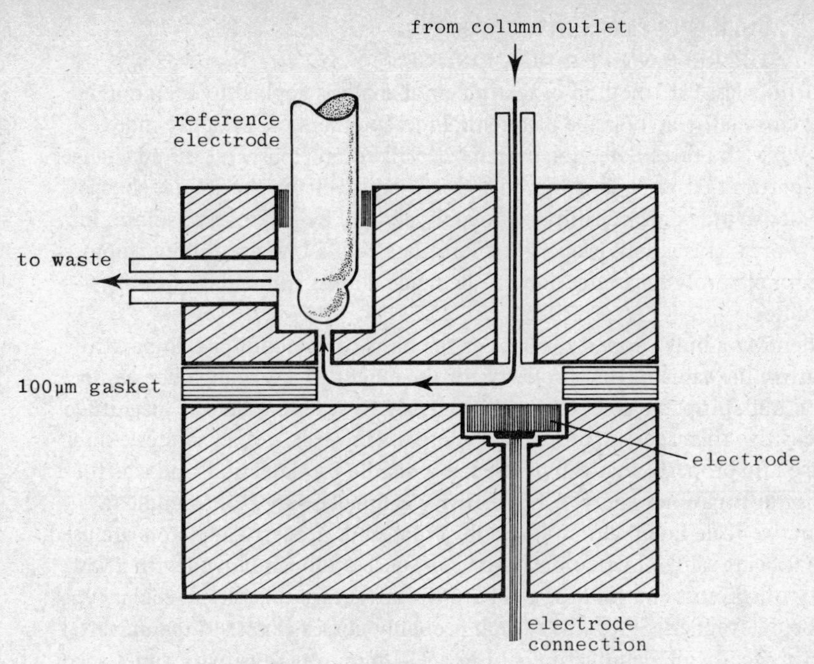

Figure 10.9. Thin liquid layer amperometric flow-cell: diameter of electrode = 2 mm; V_{cell} = 0.5 mm^3.

detectors that give a direct measure of the total amount of reacting solute, the electrochemical detectors that appear to be gaining the widest acceptance are those using the amperometric or polarographic mode. Such systems convert only a small proportion (typically 0.1–1%) of reactant molecules to products, and although such reduction of observed current might be expected to reduce the sensitivity, the sensitivities observed in practice are little worse than those of the coulometric detectors.

A major advantage of the ostensibly less sensitive amperometric system lies in the great reduction possible in the cell volumes and the very small dispersion produced by such detectors. Knox *et al.* (14) made full use of this in a detector fitted with radially movable platinum wire electrodes for their unique measurements of radial dispersion in packed columns. However, a more typical amperometric flow cell uses carbon electrodes, either highly polished glassy carbon (15) or a thick paste of colloidal carbon (16) in mineral oil, mounted in one face of a split cell block as shown in figure 10.9. Since the electrode "button" in such a design is usually 2–4 mm in diameter, V_{cell} may be as low as 1 mm^3 or less, providing minimal loss of column resolving power in the detection unit. The two commercially available amperometric detectors are of this carbon electrode type and have proved to be very satisfactory for analyses of quite a wide range of substances.

10.4 BULK PROPERTY DETECTORS: THE REFRACTIVE INDEX MONITORS

A fourth "classical" method of instrumental analysis applied to detection in liquid chromatography is the differential measurement of refractive index (RI). While the three detection systems described previously measured changes in properties that were contributed almost exclusively by the solute, changes in refractive index are contributed roughly equally by solvent and solute. In other words RI is a bulk property of solutions, while UV absorption, fluorescence, or electrolytic conduction can be made almost completely specific to the solute.

Whenever a bulk property is used as the basis of detection it is imperative to control the value of that property for the eluent to a very high degree. In general bulk property detectors will inevitably be several orders of magnitude less sensitive than specific property detectors. However, not all solutes exhibit any specific property that can be used as a handle for detection, and it is for these or in situations where high sensitivity is unnecessary (for example in preparative scale liquid chromatography) that bulk property detectors are used.

That being said, RI detectors have been successfully employed with a wide variety of solvents and solutes, and have the advantage that all molecular types may be determined. Since this system probably comes closest to the universal detector, nearly all manufacturers offer an RI monitor. Sensitivity varies according to maker, but typical models have a noise level, N_D, of 10^{-7} RI units while the most sensitive commercial model available can detect an RI difference of 10^{-8} RI units.

Since the range of refractive indices of liquids (see Appendix) is of the order of 0.1 RI units it is seen that the noise equivalent concentration corresponding to 10^{-8} RI units is around 10^{-7} g cm^{-3}. This is about 1000 times the value for a UV photometer when detecting a solute with $\epsilon = 10^4$ mol^{-1} dm^3 cm^{-1} and MW = 100. Even this sensitivity can only be achieved under extremely well-controlled experimental conditions, due to the sensitivity of RI to changes in temperature, pressure and eluent composition, and then some resolving power is generally lost. Data on the rate of change of RI with temperature and pressure are given in table 10.2.

Evidently the temperature fluctuations must be kept below about 2×10^{-5} K when using organic eluents (1×10^{-4} K when using water) while the pressure fluctuations must be held below about 2×10^{-4} bar or about 0.3 mm Hg.

Control of the eluent composition is also critical. For example, with a mixed eluent any selective evaporation of one component will produce changes that could easily result in changes of RI far in excess of 10^{-8} RI units. Thus when using an RI detector at high sensitivity it is necessary to avoid any evaporation of the eluent and to keep the eluent continually stirred. Column thermostatting is also essential. As pointed out in earlier chapters, any column packing material of high surface area will preferentially adsorb one or other component of a mixed eluent. Any fluctuation in column temperature will produce a change in this balance and a corresponding change in the composition of eluate. Rough calculation shows that to hold RI changes due to this effect below 10^{-8} RI

Table 10.2. Refractive index change with temperature and pressure.

Solvent	H_2O	C_2H_5OH	CH_2Cl_2	C_6H_{14}
RI, n	1.333	1.360	1.335	1.425
(dn/dT) (K^{-1})	1.0×10^{-4}	4×10^{-4}	6×10^{-4}	5×10^{-5}
(dn/dp) (bar^{-1})	1.3×10^{-5}	4×10^{-5}	5×10^{-5}	6×10^{-5}
stability[a] of T (K)	10×10^{-5}	2.5×10^{-5}	1.8×10^{-5}	2×10^{-5}
stability[a] of p (bar)	6×10^{-4}	2.5×10^{-4}	2×10^{-4}	1.8×10^{-4}

[a]Stability required to obtain n constant to 10^{-8} RI units.

units, the column must be thermostatted to about 10^{-3} K (17).

The relative temperatures of the eluate stream and the reference sample of eluent is generally controlled by a coiled heat exchanger through which the eluate is passed before reaching the measuring cell. The whole detector cell should itself be additionally thermostatted, as should the column and, to a lesser degree, the eluent reservoir. Any heat exchanger inevitably produces additional peak dispersion but Deininger and Halasz (18) have demonstrated an effective modification of one commercial tubing coil heat exchange system, which much reduced this additional zone broadening. However, even with this modification the RI-detector/heat-exchanger combination may still exhibit zone broadening characteristics that are not compatible with modern high-resolution chromatographic column systems.

Figure 10.10 shows a typical "thin layer" measuring cell for RI detection by measurement of changes in relative refraction at the prism-sample interface. Such measuring cells may have swept volumes, V_{cell}, of less than 5 mm^3.

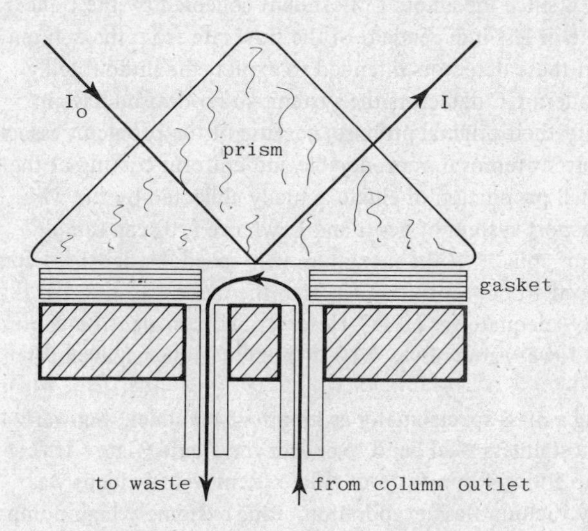

Figure 10.10. Refractive index flow-cell reference and measuring channels parallel along roof axis of prism. Monitors ΔRI by change in percentage of reflection of light beam. $V_{cell} = 5\ mm^3$.

Other principles can also be used to monitor refractive index changes in solutions, and one outstanding (and expensive) RI detector uses the interferometer technique, which effectively measures the difference in velocity of light in the two sampling channels.

10.5 TRANSPORT DETECTORS

In 1964 James, Ravenhill and Scott (19) conceived an early and entirely novel detector for liquid chromatography. It was based upon the principle of physically separating the eluent, which was necessarily volatile, from the solute, which was necessarily involatile. The principle of this form of detection is shown in figure 10.11. The eluate was collected on a moving wire, which first passed into an oven heated to 100°C where the eluent was evaporated. The wire with solute remaining on it passed into a second compartment heated to 800°C where pyrolysis took place to give gaseous products. These were then passed in a stream of nitrogen to a flame ionisation detector (FID) as used in gas chromatography. Pyrolysis proved to be a slow and irreproducible chemical process and was replaced by combustion by Scott and Lawrence (20). The CO_2 thus formed was reduced catalytically to methane and passed to the FID. Other gas-phase detection systems have been used, and probably the most important recent modifications are the use of a mass spectrometer for both quantitation and identification of solutes on the wire or band (21) and of an electron-capture detector (22).

Other transport systems have been described based upon the same general idea. These are discussed in detail by Scott (1).

While the commonly used FID is a mass-sensitive detector rather than a concentration detector, the complete transport system behaves as a concentration detector for LC since the amount of effluent collected by the transporting system is more or less independent of the flow rate from the column. While the designers of these detectors intended to exploit the undoubtedly high sensitivity of modern GC detectors the systems so far devised have in general failed to justify their original promise, because of the problems associated with selective solvent removal, reproducible and uniform coating of the wire, and the very small proportion of eluate actually collected by the wire.

Thus the wire transport system of Scott and Lawrence (20) can sample eluate at about 10 $mm^3 \ min^{-1}$ at the maximum wire speed. Under these conditions the noise equivalent concentration C_N is in the range 2 to 6×10^{-6} g cm^{-3}, which is hardly adequate for HPLC. However, with an electron-capture detector a very much lower value of $C_N = 10^{-9}$ g cm^3 can be achieved for a favourable sample.

In their work using a mass spectrometer as monitor McFadden, Schwartz and Evans (21) used a stainless steel band to collect the eluate (figure 10.11b). This collector took up about 4 $mm^3 \ s^{-1}$ of eluate. Removal of eluent was then accomplished by vacuum flash evaporation, using extremely high pumping speeds. Thus to achieve a pressure of 0.1 mm Hg in the pumping chamber using hexane as eluent a pumping speed of 600 $dm^3 \ min^{-1}$ is required. In practice this system achieves an enrichment of sample over eluent by a factor

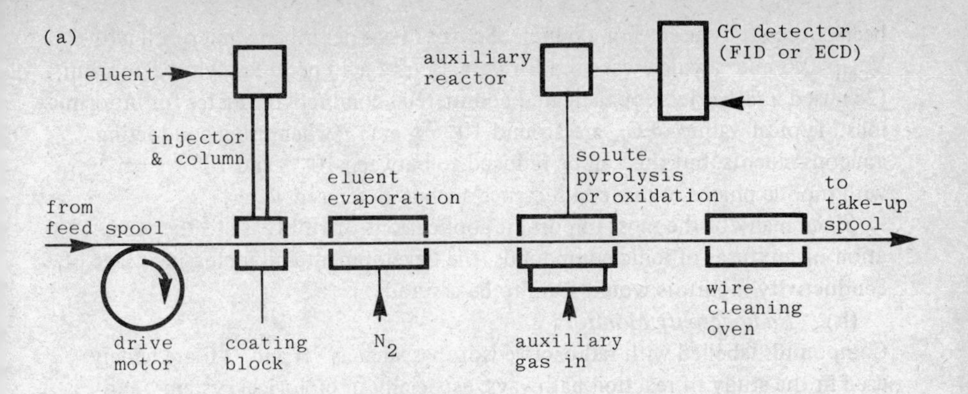

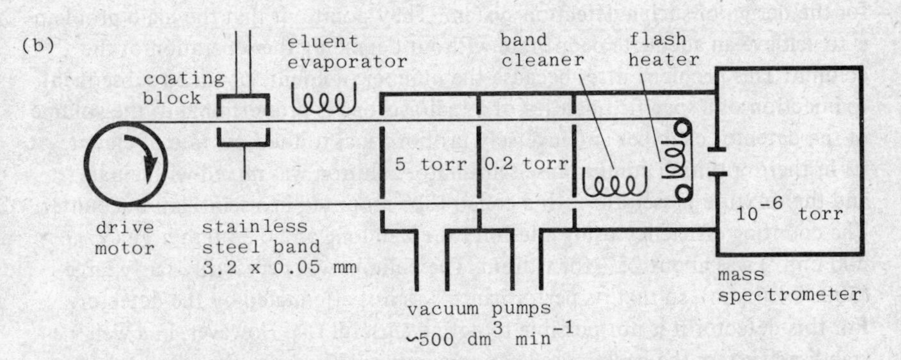

Figure 10.11. Outline diagram showing principle of transport detector: (a) wire-transport system with FID or ECD; (b) Finnigan band transport system with mass spectrometric detection.

of 10^5 with hexane and similarly volatile solvents, and can monitor solute concentrations of 10^{-7} to 10^{-8} g cm^3 by following production of a single ion at a fixed m/e ratio. This instrument does appear to be a thoroughly practical, if expensive, means of monitoring HPLC column effluents, and since the data it produces are capable of qualitative as well as quantitative interpretation further developments are confidently expected.

10.6 OTHER HPLC DETECTORS

This section covers those other less widely used monitors that have proved useful as detectors for HPLC and appear to have potential for further development.

(a) *Conductivity Monitors*

Many solutes separated by HPLC are either ionised or capable of interaction with a solvent to produce ions, and consequently a continuous record of the electrical conductivity of the effluent can be used to detect eluting solutes. Conductivity of solutions cannot be measured using direct current because of polarisation at the electrodes, but high frequency AC conductivity meters have been successfully used. For example, Scott (1) has described a microcell with $V_{cell} < 0.6\ mm^3$, which was used for detection of acid peptides, while Baumann (23) used a rather less sophisticated commercial conductivity meter for inorganic ions. Typical values of C_N are around 10^{-6} g cm^{-3} when using conducting aqueous eluents, but this can be reduced to between 10^{-8} and 10^{-9} g cm^{-3} with mobile phases of low conductance such as deionised water.

Since many of the most important applications of HPLC are to the separation of mixtures of ionic compounds, the development and increasing usage of conductivity monitors would seem to be assured.

(b) *Radioactivity Monitors*

Compounds labelled with radioactive isotopes such as ^{3}H and ^{14}C are widely used in the study of reaction pathways, especially in biological systems, and there are clearly wide potential applications for a radiometric detector for HPLC. Reeve and Crozier (24) have considered in detail the optimum parameters for the design of such a detection system. They point out that the main problem is to achieve an adequate count rate without degrading the separation of the column. This problem arises because the number of counts obtained subsequent to injection of a specific quantity of a radioisotope is proportional to the volume of the detector chamber and inversely proportional to the flow rate of eluent.

In their optimum compromise scintillator solution was mixed with eluate and the mixture passed through a coiled tube mounted in a scintillation counter. The counting efficiency using a teflon tube 2 m long and 0.35 mm bore (V_{cell} = 200 mm^3) was about 25% for tritium. The column was necessarily fairly large (500 × 4.6 mm), so that its performance was not attenuated by the detector. For this detector it is not possible to define a useful C_N. However, in a well-stabilised system the background count is around 30 counts per minute. In the case of Reeve and Crozier's experiments the noise levels represented about 2 to 6×10^{-14} mol of ^{3}H in the detected substances.

Several designs of flow scintillation counters with flow-cells packed with particulate solid scintillators have been described, but counting efficiencies of these heterogeneous systems are not as high as with the homogeneous system described above. Nevertheless, detection of amounts of ^{3}H and ^{14}C approximately one order of magnitude greater than quoted above has been achieved with the packed bed flow-cells, which should theoretically be capable of construction such that chromatographic zone broadening is comparable to that of an equivalent size separation column. Radioactivity monitors do serve a specialised requirement, and one can confidently predict their continued development.

(c) *Dielectric Constant Monitors*

Although no commercial instrument that monitors the dielectric constant of the eluate has appeared to date, the systems devised by Vespalec (25) and Klatt (26) demonstrate that detection can be achieved at concentration levels as low as 10^{-5} to 10^{-6} g cm^{-3}, very similar to those measurable by refractive index monitors. Regrettably, since dielectric constant is a bulk property, this detector suffers from the same faults as the RI monitor, namely its extreme sensitivity to temperature and eluent composition. It does, however, have the merit that the monitoring volume can be made exceedingly small. The dielectric monitor appears to offer a reasonable alternative to the RI detector.

(d) *Flame Photometric Monitors*

HPLC separations of inorganic ions and complexes are becoming increasingly important, and direct coupling of a flame photometer to the outlet from an HPLC as described by Freed (27) may well be generally applicable to detection of metal and metallo-organic samples from aqueous separation systems. Although there are no data on the zone broadening characteristics of this combination the noise equivalent concentration for sodium ion was about 10^{-7} g cm^{-3} and, for a range of other cations in the non-selective mode, about 10^{-6} g cm^{-3}. However, the combination was very flow-rate-sensitive and worked at highest efficiencies only when the flow into the burner was in the same range as the designed sample aspiration rate. Fortunately this was close to the normal flow rates used in HPLC, at between 0.7 and 0.9 cm^3 min^{-1}.

10.7 POST-COLUMN REACTORS

It is evident from the foregoing sections that there is as yet no detector for HPLC that has universally high sensitivity for all solutes, and indeed it seems highly unlikely that any such detector will ever appear. The versatility of a selective detector can, however, be greatly extended by the use of a suitable chemical reaction, which forms a derivative or product that can be more easily detected. Such reactions can be carried out either before or after passage of the solute through the HPLC column. The subject is discussed comprehensively by Lawrence and Frei (28). Pre-column derivatisation would normally be carried out prior to injection and so is not strictly related to detectors. It tends to be less than 100% quantitative, it may produce degradation of the solute of interest, and it may produce several different derivatives of a particular solute. In general post-column derivatisation is preferable since it need not be quantitative, and the material actually detected need not be a derivative of the solute but can equally well be a derivative of the added reagent or indicator. Since post-column derivatisation requires specific modifications to the detection system it is appropriately discussed in this chapter.

The classical example of a post-column reaction to enhance detectability is the ninhydrin reaction widely used in amino-acid analysers. Here a solution containing ninhydrin is mixed with the eluate from an ion-exchange column. The mixture is heated while passing through a time-delay coil in which any amino groups react to form the strongly coloured dihydrindantylamine ($\epsilon = 2 \times 10^4$ mol^{-1} dm^3 cm^{-1} at 570 nm), which is readily detected by a visible flow photometer. The system in its modern form achieves $C_N \approx 2 \times 10^{-8}$ g cm^{-3} with HPLC equipment (29) and is illustrated in figure 10.12. The time delay coil is 18 m in length and 0.25 mm in bore, the reaction temperature is 120°C and the system is pressurised so that boiling is avoided.

Any flow reactor has two major parts, a mixing section and a time-delay/reaction chamber. The performance of both of these must be optimised if unacceptable spreading of chromatographic zones is to be avoided. A definitive paper by Snyder (30) discusses aspects of design of both the mixing and reaction chambers in systems based on gas-segmented large-bore coil reactors (figure

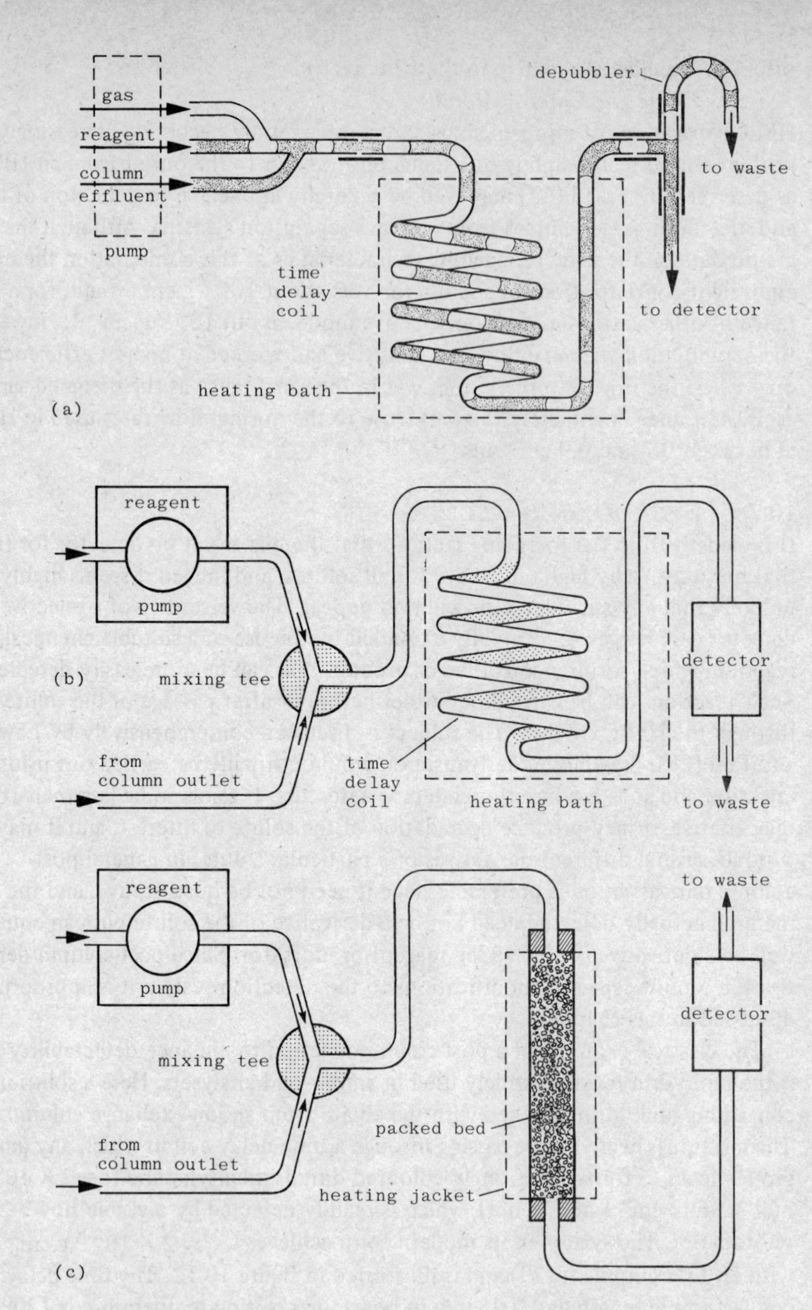

Figure 10.12. Continuous flow post-column reaction systems: (a) Gas-segmented coil, tubing bore = 1.5–2.5 mm; (b) Capillary coil, tubing bore = 0.25–0.8 mm; (c) Packed bed, for which typical packings are glass beads of d_p = 10–100 μm.

10.11a), and shows that it is possible to construct a reactor/colorimeter analyser that will not increase the zone volume of an unretained peak eluted from a 5000-plate column by more than 5% with a maximum reaction time of 8.5 minutes.

Jonker (31) and Frei *et al.* (32) have investigated the band-spreading characteristics of the other two alternative forms of time-delay/reaction chamber; the capillary coil (figure 10.12b) and the packed bed (figure 10.12c) and their associated mixing systems. Both groups of workers have produced reactor designs that increase the zone volume of unretained peaks from a 4000-plate, 250 × 3 mm column by 50% (31) for a 70 s reaction time and 50% (32) for a 30 s reaction time. All the above systems are operated in such a way that only one reagent stream is mixed with the column effluent in the reactor; more complicated reaction flow systems have not yet been used in HPLC experiments.

The reaction/detection combinations described above use optical means of sensing reaction products; photometry (29, 31) and fluorimetry (32). However, post-column reactions may also be used to generate electroactive species, which may be detected by an electrochemical monitor. Takata and Muto (33) devised such an electrochemical flow reactor in which ferricyanide reacted at 80°C with reducing sugars to give ferrocyanide, which was detected electrochemically. Since they used a flat packed bed of large volume (600 mm^3) the band dispersion produced by this device was far too large for HPLC. Nevertheless, with attention to design there is no reason why a reactor coupled to electrochemical detection cannot be just as effective as the more usual optical combinations.

The last reactor/detector assembly that will be discussed here is one that uses nephelometric or light-scattering sensing of separated solutes after admixture of the column effluent with a liquid that precipitates the solutes as solid or liquid colloidal particles. Novotny *et al.* (33) precipitated non-polar lipids from a 2:1 (v/v) acetone/methanol solvent by addition of aqueous ammonium sulphate solution via a mixing tee and a short capillary-tubing mixing coil, and using laser illumination they determined the concentration of the precipitated solutes by measuring the scattered light from the flow-cell, as shown in figure 10.13. From the published chromatograms a value of C_N of about 5×10^{-7} g cm^{-3} can be derived. Unfortunately the detector response is non-linear, but this instrument may well be capable of development for use in fields where good detection systems are at a premium.

In conclusion it must be emphasised that each reactor must be tailored to its particular end use. For every post-column reaction there will be a set of optimum operating parameters such as temperature, concentration of reagent, flow rate of reagent, reaction time, etc. These variables have to be studied in detail in order to optimise the configuration and operating conditions of the reactor. Thus each final reactor design will tend to be specific for a single end use, and it may well be limited to a small range of eluent flow rates.

The time needed to determine the relevant variables and hence to optimise reactor design is quite substantial, but well-engineered systems can be made to operate reliably and in an analytically satisfactory fashion. Reaction/detector

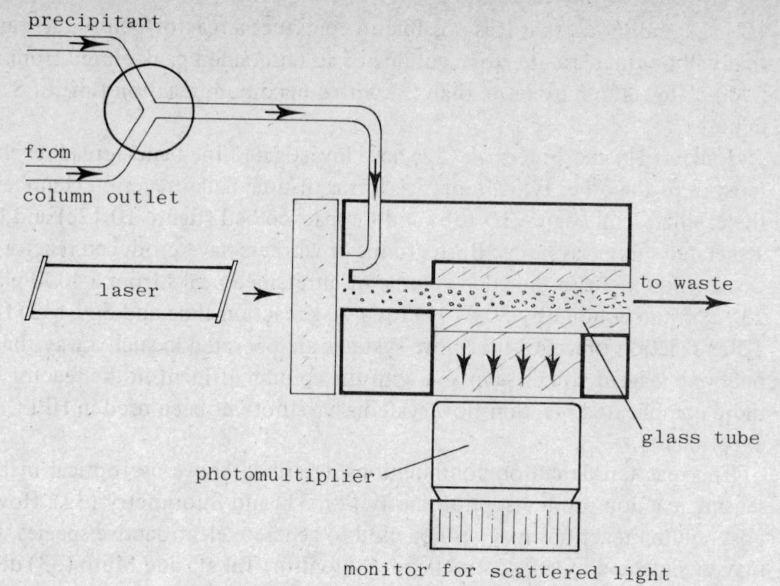

Figure 10.13. Light-scattering monitor: measuring tube length = 16 mm; tube bore = 1.2 mm; V_{cell} = 20 mm^3.

combinations will certainly find wider HPLC usage as instruments dedicated to a single analytical problem are developed using modern column technology.

10.8 PEAK DISPERSION BY DETECTORS

While the criteria (a) to (e) of section 10.2 refer to the geometrical configuration of the detector cell and to joint properties of the sensor and eluate, only (f) and (g) relate to the dynamic relationship of the column and detector. In considering the selection of the best type of detector for any application this last relationship is crucial, for a detector of poor geometry may completely or at least very seriously attenuate the separation achieved by the column.

Table 10.1 does not list values of the dispersion of different detectors since this can vary widely from one commercial model to another, even for the same type of detector, and depends quite strongly upon the operating conditions. The dispersion due to the detector and any ancillary tubing should always be determined experimentally on the actual instrument. It is obtained by making an injection of below 1 mm^3 into an injection head connected directly to the detector. The experiment should be carried out at several volumetric flow rates, bracketing those likely to be used in practice. The result of such an experiment using a "zero-length column" is illustrated in figure 10.14.

The standard deviation, σ_V, which is one-quarter of the base peak width w_{det} for a Gaussian peak, can be obtained in a number of ways, as discussed by Kirkland *et al.* (35), but in practice it is adequate in the present case to draw tangents at the points of inflection of the recorded peak and to take w_{det} as the volume of eluent corresponding to their intersection with the baseline. From figure 10.14 it is seen that w_{det} = 50 mm^3. This is probably

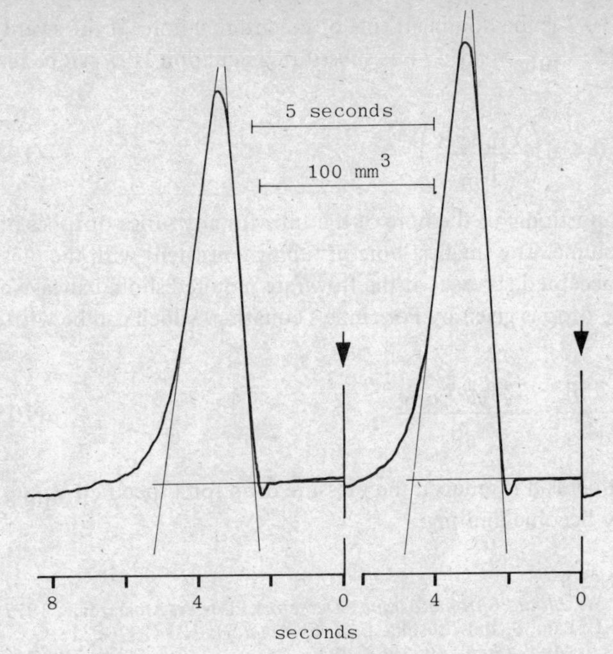

Figure 10.14. Replicate recorder traces obtained from injection of 3 mm^3 of a UV-absorbing solution directly into a UV detector. Cell compartment 10 x 1 mm; inlet tubing 110 x 0.125 mm; flow rate 20.8 $mm^3\ s^{-1}$; chart speed 10 mm s^{-1}. Peak width at base 2.45 s or 51 mm^3.

typical for a fairly well-designed injector/detector system and it is only slightly greater than the value expected for a fully stirred mixing chamber with a volume of 8 mm^3, namely $4V_{cell} = 32$ mm^3 (36).

When heat exchangers or reactor coils are necessary prior to the detector cell proper, additional dispersion will result. For laminar flow in a round tube of bore d and length L this is given by

$$w_{tube} = 0.36\, d^2 \left(\frac{Lf_v}{D_m}\right)^{1/2} \tag{10.7}$$

as discussed in chapter 8. In practice the narrow-bore stainless steel tubing used for connections and heat exchangers is relatively rough internally, and the flow is not truly laminar. Consequently, as Scott and Kucera have shown experimentally (37), the dispersion is less than that given above. On the other hand, with the smooth-walled plastic tubing commonly used for reactors, equation 10.7 is likely to be accurate. It would therefore be wise to regard equation 10.7 as providing an upper limit for the dispersion when considering the design of heat exchanger reaction coils, etc. It is interesting to note that at flow rates of 1 $cm^3\ min^{-1}$ the coil used in Reeve and Crozier's radioactivity monitor (24) would produce $w_{tube} = 85$ mm^3, while the long reactor coil used by Kraak *et al.* (29) would produce $w_{tube} \approx 300$ mm^3. The most important factor to note in equation 10.7 is the dominant role of the internal bore. If the swept volume of the tube, $V_{tube} = \pi d^2 L/4$, is substituted equation 10.7 can be recast to give

$$w_{tube} = 0.41d\left(\frac{V_{tube}\, f_v}{D_m}\right)^{1/2} \qquad (10.8)$$

Thus w_{tube} is proportional to the bore of the tube for any series of tubes of constant swept volume. The smallest bore of tubing consistent with the pressure drop that can be accepted across it at the flow rate required should always be used. The pressure drop is given by Poiseuille's equation, which can be written in the form

$$\Delta p = \frac{41\, f_v \eta L}{d^4} = \frac{52\, f_v \eta\, V_{tube}}{d^6} \qquad (10.9)$$

It is readily seen that as d is reduced the pressure drop for a specified V_{tube} and f_v can quickly become limiting.

REFERENCES

1. Scott, R.P.W., *Liquid Chromatography Detectors.* Elsevier, Amsterdam, 1977.
2. Milano, M.J., Lam, S. and Grushka, E., *J. Chromatogr. 125* (1976) 315.
3. Kirkland, J.J., *Anal. Chem. 40* (1968) 391.
4. Little, J.N. and Fallick, G.J., *J. Chromatogr. 112* (1975) 389.
5. Huber, J.F.K., *J. Chromatogr. Sci. 7* (1969) 172.
6. Brooker, G., *Anal. Chem. 43* (1971) 1095.
7. Jadamec, J.R., Saner, W.A. and Talmi, Y., *Anal. Chem. 49* (1977) 1316.
8. Sepaniak, M.J. and Young, E.S., *Anal. Chem. 49* (1977) 1554.
9. Koen, J.G., Huber, J.F.K., Poppe, H. and den Boef, G., *J. Chromatogr. Sci. 8* (1970) 192.
10. Vermula, W., *Roczniki Chem. 26* (1952) 281.
11. Kissinger, P.T., *Anal. Chem. 49* (1977) 447A.
12. Lankelma, J. and Poppe, H., *J. Chromatogr. 125* (1976) 375.
13. Buchta, R.C. and Papa, L.J., *J. Chromatogr. Sci. 14* (1976) 213.
14. Knox, J.H., Laird, G.R. and Raven, P.A., *J. Chromatogr. 122* (1976) 129.
15. Fleet, B. and Little, C.J., *J. Chromatogr. Sci. 12* (1974) 747.
16. Kissinger, P.T., Refshauge, C., Dreiling, R. and Adams, R.N., *Analyt. Letters* 6 (1973) 465.
17. Knox, J.H., *UV Spectroscopy Group Bulletin* (1977) Supplement 5, p.2.
18. Deininger, G. and Halasz, I., *J. Chromatogr. Sci. 8* (1970) 499.
19. James, A.T., Ravenhill, J.R. and Scott, R.P.W., *Chem. and Ind.* (1964) 746.
20. Scott, R.P.W. and Lawrence, J.G., *J. Chromatogr. Sci. 8* (1970) 65.
21. McFadden, W.M., Schwartz, H.L. and Evans, S., *J. Chromatogr. 122* (1976) 389.
22. Chamberlain, A.T. and Marlow, J.S., *J. Chromatogr. Sci. 15* (1977) 29.
23. Small, H., Stevens, T.S. and Baumann, W.C., *Anal. Chem. 47* (1975) 1801.
24. Reeve, D.R. and Crozier, A., *J. Chromatogr. 137* (1977) 271.
25. Vespalec, R. and Hana, K., *J. Chromatogr. 65* (1972) 53.
26. Klatt, L.N., *Anal. Chem. 48* (1976) 1845.
27. Freed, D.J., *Anal. Chem. 47* (1975) 186.
28. Lawrence, J.F. and Frei, R.W., *Chemical Derivatisation in Liquid Chromatography.* Elsevier, Amsterdam, 1976.
29. Kraak, J.C., Jonker, K.M. and Huber, J.F.K., *J. Chromatogr. 142* (1977) 671.
30. Snyder, L.R., *J. Chromatogr. 125* (1976) 287.
31. Jonker, K.M. (Doctoral Dissertation) *Post-Column Reaction Detection and its Application to the Analysis of Amino Acids.* University of Amsterdam, 1977.
32. Frei, R.W., Mickel, L. and Santi, W., *J. Chromatogr. 142* (1977) 261.

33. Takata, Y. and Muto, G., *Anal. Chem. 45* (1973) 1864.
34. Jorgenson, J.W., Smith, S.L. and Novotny, M., *J. Chromatogr. 142* (1977) 233.
35. Kirkland, J.J., Yau, W.W., Stoklosa, H.J. and Dilks, C.H., *J. Chromatogr. Sci. 15* (1977) 303.
36. Knox, J.H., *J. Chromatogr. Sci. 15* (1977) 352.
37. Scott, R.P.W. and Kucera, P., *J. Chromatogr. Sci. 9* (1971) 641.

11. APPLICATIONS OF HPLC IN CLINICAL, PHARMACEUTICAL AND INDUSTRIAL RESEARCH

An extensive applications literature has been published and several pertinent reviews on the use of HPLC have appeared in the areas of pharmaceutical analysis (1–3), clinical chemistry (4), forensic science (5) and toxicology (6). A few books specifically dealing with applications of HPLC have been published (7–9) but these may now be slightly out of date. A recent comprehensive review of the applications literature has now been completed (10). This chapter does not attempt to cover the current literature comprehensively but rather to illustrate the usefulness of the technique in diverse fields.

11.1 COMPARISON WITH OTHER METHODS OF ANALYSIS

A comparison of HPLC with other widely used analytical methods is given in table 11.1. Spectroscopic methods (UV/visible, fluorescence) are fairly non-specific and will not usually discriminate between closely related metabolites or degradation products. Fairly lengthy clean-up procedures are also necessary to minimise background absorption from endogenous materials. Radioimmuno-assay methods require expertise to raise specific antibodies but frequently nowadays kits are commercially available. However, although good sensitivity is obtained the technique often lacks specificity.

Thin-layer chromatographic methods are widely used and inexpensive. However, quantitative analysis is difficult, although TLC has the advantage that several assays can be performed simultaneously. Recently high-performance TLC systems have been developed (11), which may overcome some of the disadvantages of this technique.

Gas chromatography shows good specificity and sensitivity. However, the elevated operating temperatures used may occasionally cause thermal degradation of compounds, and it is often necessary to derivatise samples to increase their volatility and improve their chromatographic behaviour. The gas chromatographic analysis of substituents in body fluids frequently necessitates considerable sample clean-up (normally solvent extraction) to prevent interference from endogenous substances. Although such difficulties may often be minimised by the use of specific detection systems, some degree of pre-analytical purification is almost invariably required. Mass spectrometric detection affords very high specificity, which may enable still further reduction in sample preparation, but such systems are expensive to install and maintain, and require skilled operators. Gas chromatographic methods are generally inapplicable for the analysis of highly polar substances or of compounds with high molecular weights.

Table 11.1. Comparison of HPLC with other analytical techniques

Technique	Sample recovery	Speed (min)	Specificity	Limit of detection	Applicability	Sample preparation	Ease of operation	Cost (£)
UV/vis	yes	1–2	*	1 ng	fairly wide	clean up necessary	easy	3000–5000
Fluorimetry	yes	1–2	*	100 ng	restricted	”	”	”
TLC	yes	30	**	20 ng	very wide	minimal clean up	”	50–100
GLC								
a) FID			**	50 ng	wide	considerable clean up	”	3,000
b) ECD	no	10–30	***	10 pg	restricted	some clean up	”	3,500
c) N/P/S			***	1 ng	”	”	”	3,500
GC-MS			*****	100 pg	wide	”	requires skilled operator	40,000–100,000
Radioimmunoassay	no	60	**	100 pg	restricted	none	variable	5,000
HPLC								
a) UV/vis	yes		****	1 ng	wide	minimal	easy	2,500–6,000
b) Fluorimetry	yes	10–30	****	100 pg	fairly wide	”	”	3,000–6,000
c) Transport FID			**	1 μg	wide	”	”	3,500
d) E.I.	yes		*	1 μg	”	”	”	3,000–6,000

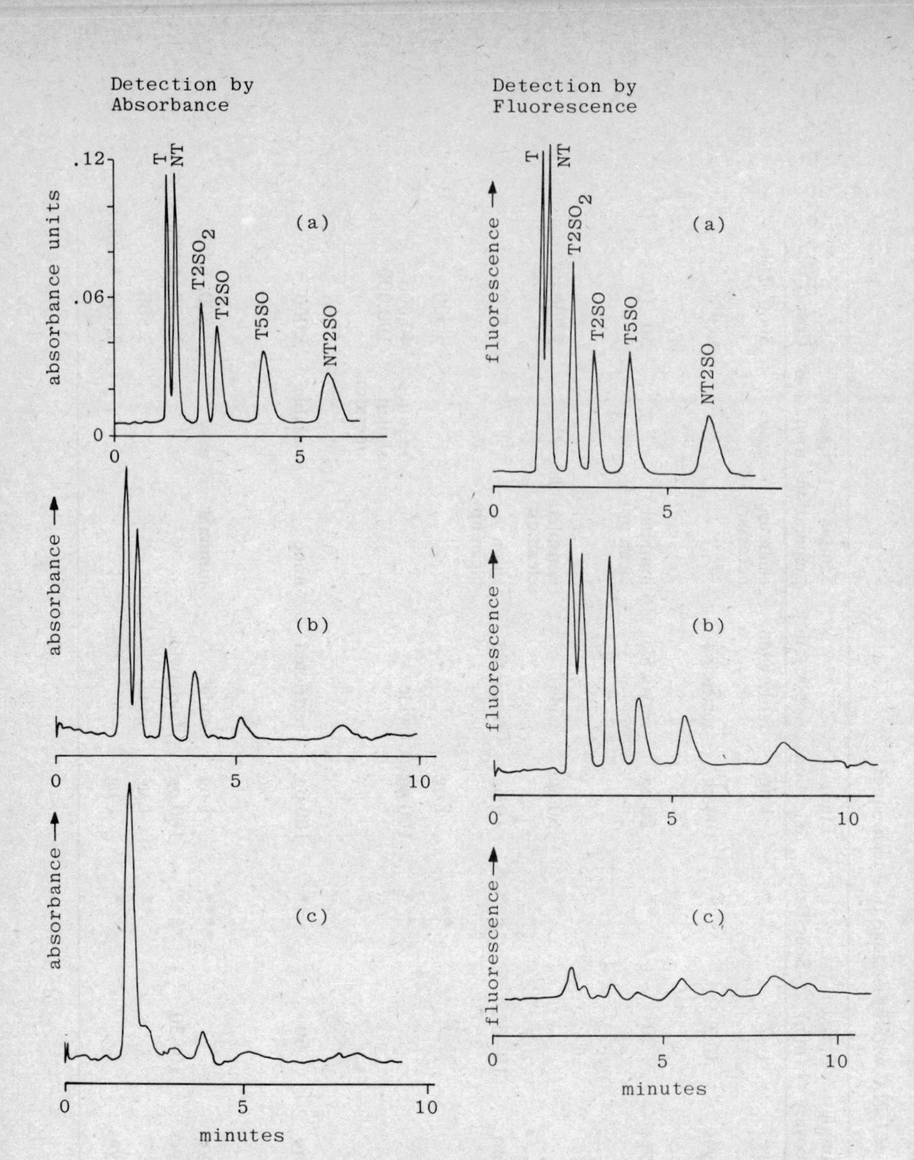

Figure 11.1. Analysis of thioridazine and metabolites using UV and fluorescence detectors after fluorogenic reaction (39).

Column: 250 x 2.8 mm
Packing: 9 μm silica gel (50 m^2 g^{-1})
Eluent: 2,2,4-trimethylpentane/2-aminopropane/acetonitrile/ethanol (95.9:1.0: 27:0.5 v/v/v/v)
Samples: (a) standard mixture of thioridazine (T), northioridazine (NT), thioridazine-2-sulphone ($T2SO_2$), T-2-sulphoxide (TSO), T-5-sulphoxide (T5SO), NT-2-sulphoxide (NT2SO)
(b) extract of 1 cm^3 plasma containing 1 μg cm^{-3} of each component
(c) extract of blank plasma
Detection: UV photometer, 254 nm, 0.12 AUFS
fluorescence excitation 365 nm, detection 440 nm

HPLC offers many advantages for quantitative analysis. Extraction procedures and sample clean-up prior to injection are much simpler than for most other methods, especially when reversed-phase HPLC systems are employed. Thus in the analysis of pharmaceutical products it is frequently sufficient to crush or mix the product with a suitable solvent, filter and inject (12–14). Body fluids may also be directly injected onto reversed-phase columns, permitting rapid routine clinical assays to be developed (15–18). This latter procedure is not recommended with adsorption systems, because the build-up of highly polar material on the top of the column results in a gradual deterioration of column performance and eventual blockage of the column. This problem may be overcome to some extent by the use of a removable pre-column, but a reversed-phase system is nearly always more suitable.

11.2 DETECTION SYSTEMS

The main limitation of analysis by HPLC lies in the detection system. The most commonly used detectors are ultraviolet spectrophotometers, preferably with variable wavelength facilities. However, many compounds of interest do not contain UV-absorbing chromophores and, therefore, cannot be detected in this way. Refractive index and moving-wire flame-ionisation detectors are fairly universally applicable, but lack the sensitivity necessary for clinical applications. This problem can often be circumvented by the preparation of suitable UV-absorbing derivatives (19), and this approach has frequently been used, as indicated in table 11.2.

Table 11.2 UV-absorbing derivatives

Compound class	Derivative	Ref.
17-keto-steroids	2,4-dinitrophenyl hydrazine (DNP)	20, 21
hydroxysteroids	benzoyl ester	22
fatty acids	phenacyl ester	23, 24
	p-bromophenacyl ester	25
	naphthacyl ester	26
prostaglandins	p-nitrophenacyl ester	27
	p-bromophenacyl ester	28
amino acids	phenylthiohydantoin (PTH)	29–36

Increased sensitivity should be available with the use of fluorimetric detectors for compounds that are naturally fluorescent, or for those that have been reacted to form fluorescent derivatives (19, 37, 38). However, such sensitivity increases are not always observed, because of limitations in the equipment. For example, in the analysis of the psychotropic drug thioridazine and its five metabolites in blood, Muusze and Huber (39) obtained similar detection limits for both UV and fluorescence detection (figure 11.1). The fluorophores were produced by post-column oxidation of the phenothiazines with excess potassium permanganate/

acetic acid in a 1.4 m length of 0.4-mm-bore tubing (reaction time 5 s). The excess permanganate was destroyed with hydrogen peroxide in 100 mm of the same tubing. The total volume of both reactors including mixing heads was 200 mm^3, the volume being kept as low as possible to avoid loss of resolution. Although the fluorimetric detector did not give increased sensitivity, the increase in selectivity was found to be useful. As shown in figure 11.1, the thioridazine UV-absorption peak is masked by a UV-absorbing constituent from the blood plasma. This interference is greatly reduced by using the fluorescence reaction detector. However, it is also worth noting that the resolution between thioridazine and northioridazine has deteriorated with the fluorimetric detector due to the peak-broadening effect of the post-column reactor.

Increased sensitivity can, however, be obtained with carefully controlled reaction conditions. Figure 11.2 shows the analysis of the nonapeptide lysine^8-vasopressin in a pharmaceutical dosage form (40). Fluorescence detection is obtained by post-column reaction with fluorescamine. As shown in figure 11.2,

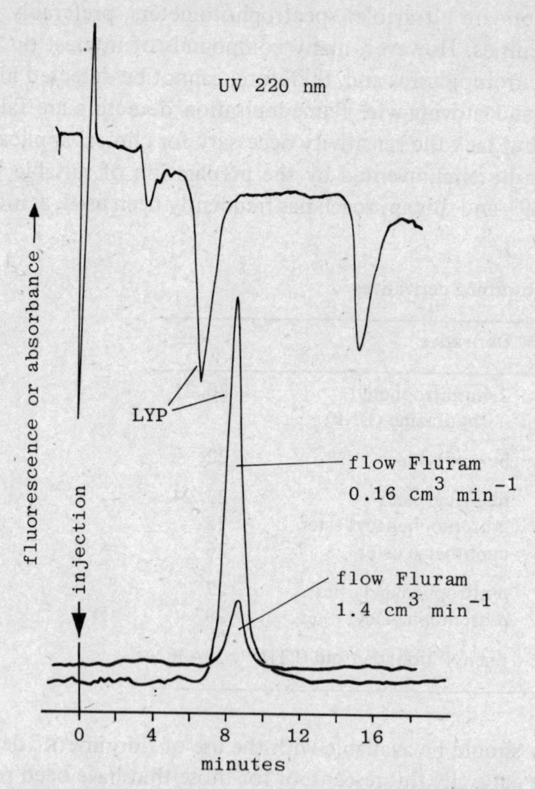

Figure 11.2. Analysis of lysine^8-vasopressin (LVP) using fluorescence for detection showing effect of flow rate of fluorogenic reagent (40).

Column: 250 x 3 mm
Packing: 10 μm Merckosorb RP8
Eluent: phosphate buffer pH 7/acetonitrile (85:15 v/v)
Detector: UV photometer, 220 nm
fluorescence

the flow rate of the reagent solution has a considerable influence on the detection signal. An increase in the flow rate from 0.16 to 1.4 cm^3 min^{-1} results in an eight-fold reduction in signal. However, at optimal reaction conditions an increase in sensitivity over the UV detection system is observed.

Several other specific detection systems have been devised, including electrochemical detectors (41–43) and electron capture detectors (44,45), which show increased sensitivity but lack general applicability. Various approaches to obtaining a suitable HPLC-MS interface have also been described (46–51). These systems are not, however, in general use, and at present they restrict the use of solutes and mobile phases. Much more research is needed before they will be of use for routine assays.

11.3 PHARMACEUTICAL AND INDUSTRIAL APPLICATIONS

The quantitation of components is usually achieved by measuring peak areas or peak heights relative to a suitable internal standard. The measurement of peak areas is inherently the more accurate of the two methods, but since peak areas are more affected by flow rate variations than peak heights (52,53) it is essential to use a pump that is capable of delivering a constant flow rate. Provided the calibration plot of peak height or peak area *vs* amount injected is linear over the required range, either method is satisfactory.

A particularly difficult problem encountered in the pharmaceutical industry is the rapid, routine analysis of multicomponent dosage forms for quality control purposes. HPLC has been found to be very useful in this area. The separation of the components present in an analgesic product is shown in figure 11.3 (54); p-chlorocetanilide is included as the internal standard. Excellent resolution of all components is achieved on a 300 mm column of μ-Bondapak-C_{18} is less than 10 min. The mobile phase consists of 0.01% aqueous ammonium carbonate/acetonitrile (60:40). The ammonium carbonate was used to maintain the pH of the mobile phase at 7.8, since the molar absorptivity of the butalbarbitone anion is greater than that of the neutral molecule. The concentrations of the active constituents were determined by peak area with good precision. No loss in resolution was observed even after 1,000 injections.

Similar procedures are, of course, applicable to the determination of unreacted intermediates, or of by-products in various industrial products. Tartrazine, a sulphonic acid dye used extensively in the food, drug and cosmetic industries, is synthetised by reaction of sulphanilic acid with 3-carboxy-1 (4-sulphophenyl)-5-pyrazolone (pyrazolone T). Determination of these impurities in dye samples was achieved by a reversed-phase soap chromatographic method (55). Rapid analysis was obtained on a column of μ-Bondapak-C_{18} with a mobile phase of water/methanol/formic acid (400:400:1) containing 3.0×10^{-3} M tetrabutylammonium hydroxide and 6.0×10^{-5} M tridecylamine (figure 11.4). In this case, since the intermediates are present at fairly low levels in the dye, two separate internal standards were included. Normal-phase soap chromatography has also been used for the analysis of impurities in dyestuffs (56). The separation of the components in sunset yellow on a column of silica was shown in figure 6.10. The eluent consisted of propanol/methylene chloride/water (70:40:12)

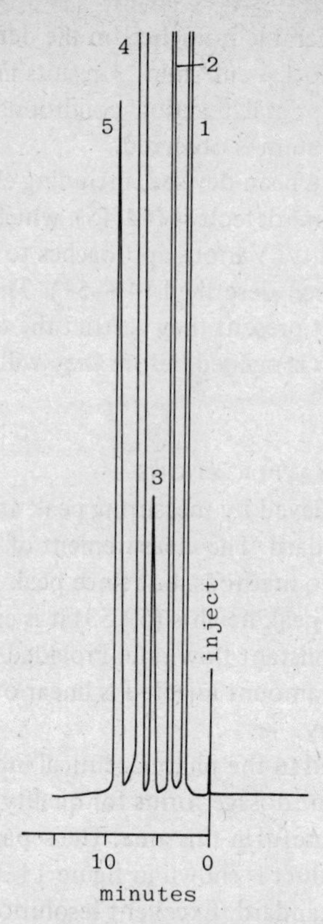

Figure 11.3. Chromatogram of analgesic mixture (54).
Column: 300 x 4 mm
Packing: 10 μm μBondapak C_{18}
Eluent: 0.01% aqueous ammonium carbonate/acetonitrile (60:40 v/v)
Solutes: (1) aspirin, (2) caffeine, (3) butalbital, (4) phenacetin, (5) p-chloroacetanilide

containing 2% w/v cetrimide (cetyltrimethylammonium bromide). These types of quality-control assays are ideally suited for inclusion in automated processes.

Assays of this type can also be used for pharmacokinetic studies. For example, a sensitive, specific procedure was developed for the simultaneous determination of phenylbutazone and its metabolite, oxyphenbutazone (figure 11.5) in plasma (51). Aliquots of plasma extracts containing the 2,4-DNP of 3,4-dimethoxybenzaldehyde as an internal standard, were chromatographed on an adsorption column using a mobile phase of hexane/tetrahydrofuran (77:23) at 35°C. The peak shape was improved by the addition of a small amount (0.002%) of acetic acid to the mobile phase. Chromatographic analysis was achieved in 8 min. UV detection at 254 nm allowed quantitative analysis of 1 cm^3 plasma samples containing less than 0.25 $\mu g\ cm^{-3}$ of phenylbutazone or oxyphenbutazone. Plasma levels of a male volunteer who had been given a 400 mg dose of phenylbutazone were monitored over 388 h (figure 11.6). A good correlation was found between the results obtained by this method and those obtained by a gas chromatographic

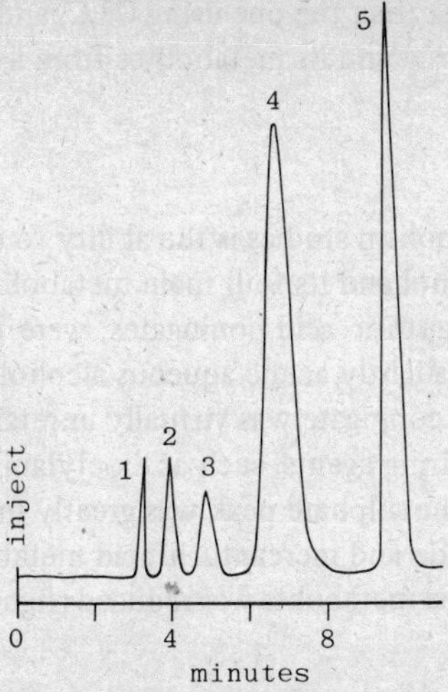

Figure 11.4. Intermediates from synthesis of tartrazine (55).

Column: 300 x 4 mm
Packing: 10 μm μBondapak C18
Eluent: water/methanol/formic acid (400:400:1 v/v/v) containing 3 x 10^{-3} M tetrabutylammonium hydroxide and 6 x 10^{-5} M tridecylamine
Solutes: (1) sulphanilic acid, (2) pyrazoline T, (3) 3-nitro-salicyclic acid, (4) tartrazine, (5) m-chlorobenzoic acid
Detector: UV photometer, 280 nm

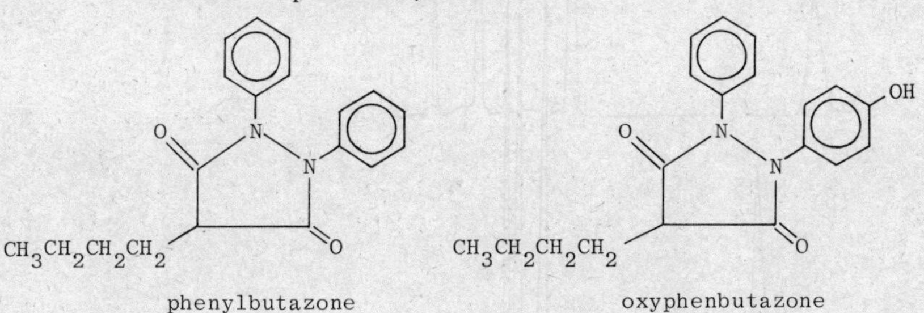

Figure 11.5. Structures of phenylbutazone and oxyphenbutazone.

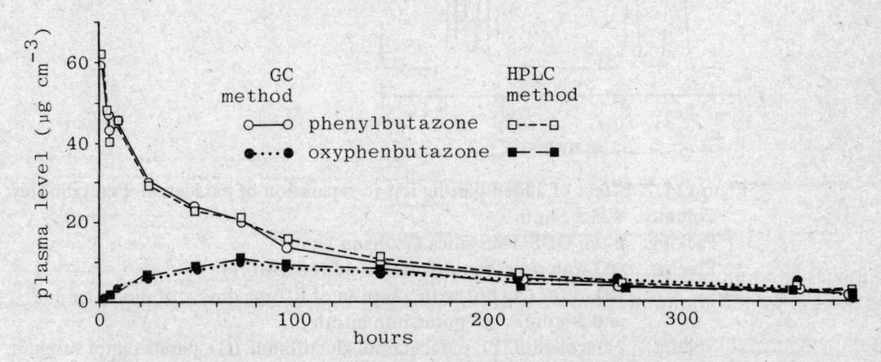

Figure 11.6. Comparison of plasma drug levels determined by HPLC and GC following a single oral dose of 400 mg phenylbutazone to a volunteer (58).

method (58). The present method is simpler than the one using GLC, and permits simultaneous determination of both the drug and its metabolites from a single plasma extract.

11.4 CLINICAL APPLICATIONS

A major advantage of HPLC in drug-metabolism studies is the ability to analyse conjugated metabolites directly. Paracetamol and its four main metabolites, the sulphate, glucuronide, cysteine and mercapturic acid conjugates, were readily separated on "capped"-ODS silica using a slightly acidic aqueous alcoholic eluent (18). Under these conditions the sulphate conjugate was virtually unretained (see figure 5.6). On addition of suitable ion-pairing agents, such as dioctylammonium or tetrabutylammonium, the retention of the sulphate peak was greatly increased (59). The retention times of the glucuronide and mercapturic acid metabolites were also increased while that of the cysteine metabolite was reduced (figure 11.7).

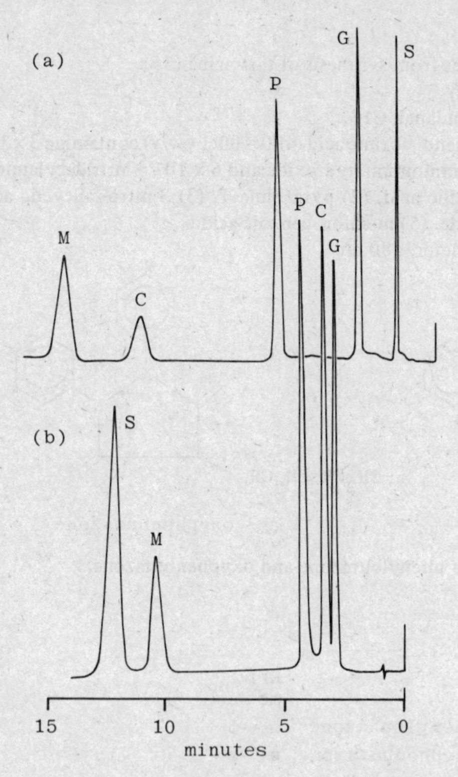

Figure 11.7. Effect of added pairing ion to separation of paracetamol metabolites.
Column: 125 x 5 mm
Packing: 6 μm ODS/TMS silica (Wolfson Unit)
Eluent: (A) water/methanol/formic acid (86:14:0.1)
(B) as for (A) but with addition of 0.7 mg dm^{-3} of dioctylamine and 3 g dm^{-3} of potassium nitrate
Solutes: paracetamol (P), paracetamol glucuronide (G), paracetamol sulphate (S), paracetamol cysteine (C) paracetamol mercapturic acid (M)
Detector: UV photometer, 242 nm, 0.2 AUFS

Excessive retention of the compounds, in particular the sulphate, could be controlled by the addition of a co-ion such as nitrate. The use of these additives in the eluent allowed a fine tuning of the retention times of the drug and its metabolites to avoid any interference from normal urinary constituents. Application of the technique to the analysis of therapeutic and overdose urines showed the presence of at least three metabolites in addition to those already known. An example of this application is shown in figure 6.9.

In routine assay situations it is desirable that the analysis should be performed using isocratic conditions. The full versatility of possible mobile-phase compositions should be investigated before resorting to gradient elution systems, since the time required for column reconditioning after each run reduces the output

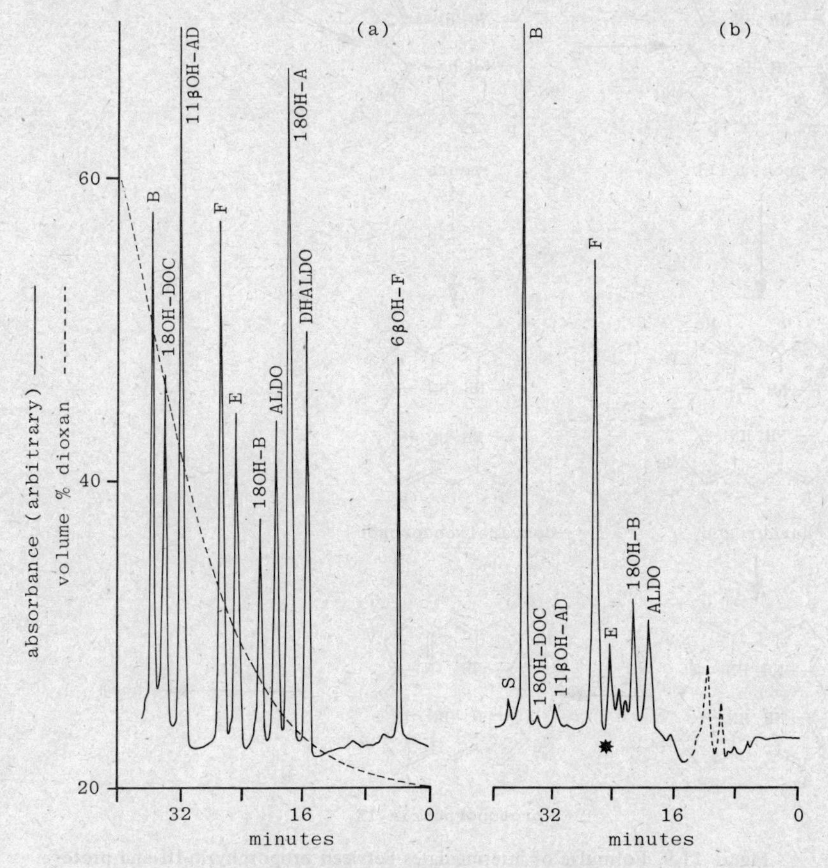

Figure 11.8. Separation of polar adrenal steroids (60): (a) standard mixture, (b) steroids secreted by an aldosteronoma (Conn's syndrome) in monolayer culture.

Column: 250 x 2.1 mm
Packing: Zorbax ODS
Eluent: gradient of dioxan/water, composition shown
Solutes: for identification see original paper
Pressure: 170 bar
Detector: UV photometer, 254 nm, 0.04 AUFS and 0.64 AUFS

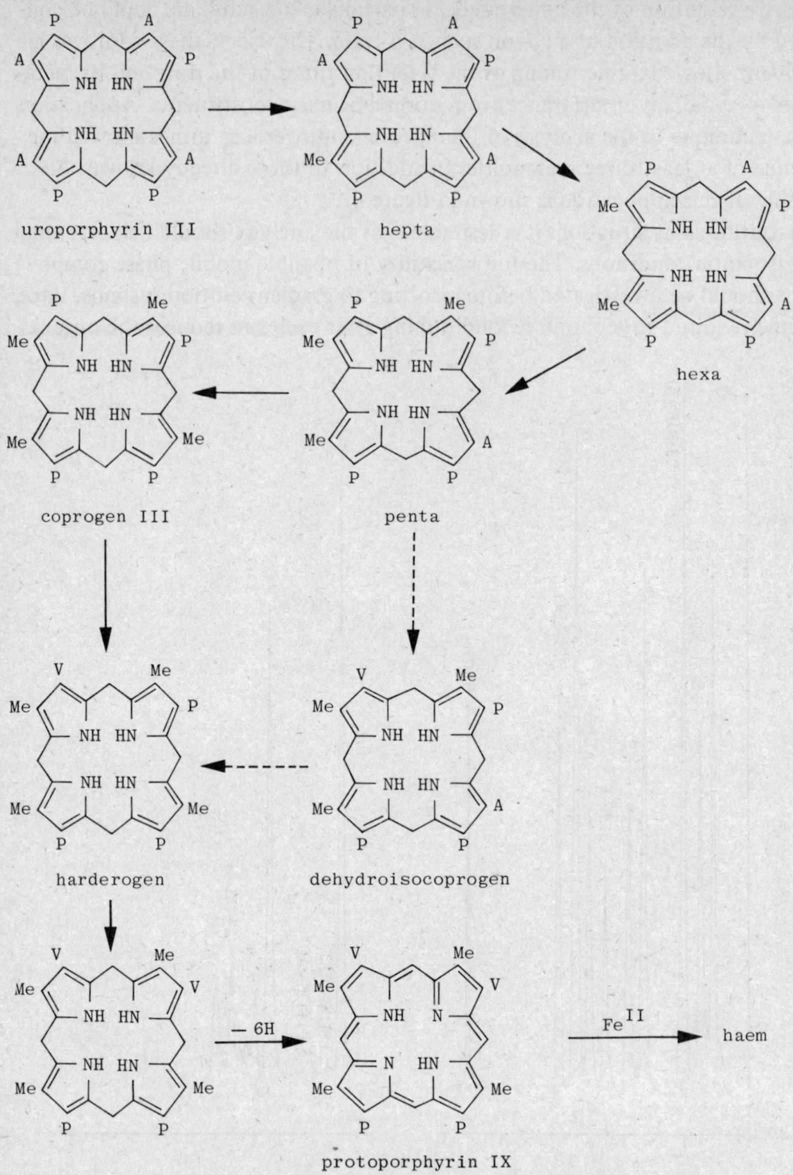

Figure 11.9. Formulae of intermediates between uroporphyrin-III and protoporphyrin-IX in the biosynthesis of haem.

of data. Nevertheless, for the analysis of complex mixtures gradient elution is sometimes unavoidable. An example of the excellent results that can be obtained is shown in figure 11.8 (60). These very polar adrenal steroids are difficult to analyse by GLC since some of them are thermally unstable even after derivatisation. However, the standards were well resolved on a column of Zorbax-ODS at 45°C using a dioxan/water gradient. The method was used to examine them

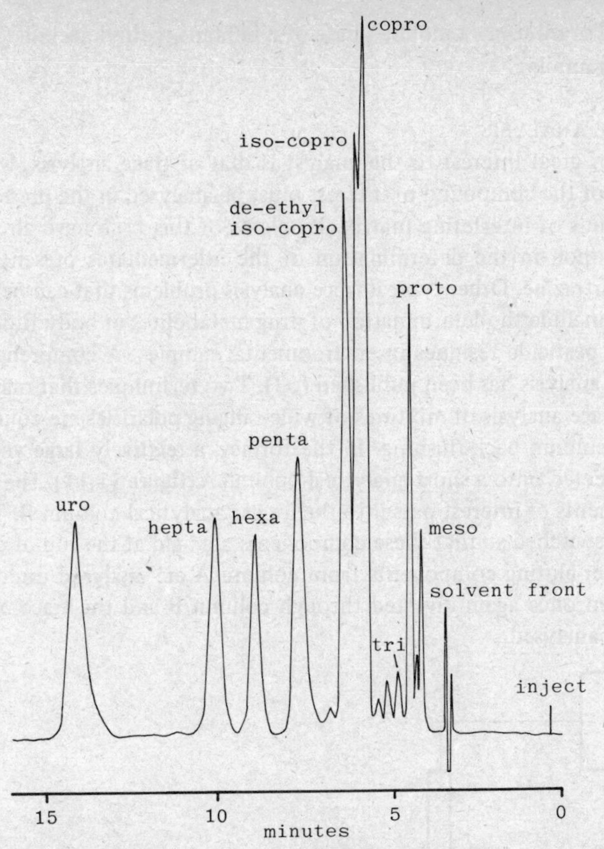

Figure 11.10. Faecal porphyrin profile from a patient with symptomatic porphyria (62).

Column: 300 x 4 mm
Packing: μ Porasil
Eluent: heptane/methyl acetate (60:40 v/v)
Solutes: see figure 11.9 for representative formulae
Detector: UV photometer, 404 nm

steroids secreted by a cultured aldosteronoma (Conn's syndrome). Levels of steroid measured from the cultures were usually well above the limit of sensitivity (2 ng) for UV-absorbing steroids. Studies such as this can be used to produce "fingerprint" patterns of steroid production from normal and neoplastic cells and hence to help in the diagnosis of disease.

Similarly, diagnosis of disorders in porphyrin metabolism necessitate the routine identification and determination of porphyrins in blood, urine and faeces (figure 11.9). Urinary porphyrins alone are not definitive with respect to the porphyrias, but consideration of total urinary porphyrins and the pattern of the faecal porphyrins allows differentiation of the diseases. The porphyrins may be analysed as their methyl esters by adsorption chromatography (61, 62). The excretion pattern of faecal porphyrins in a patient with symptomatic porphyria is shown in figure 11.10 (62). This chromatogram was obtained on a column of μ-Porasil using a mobile phase of n-heptane-methyl acetate (3:2) with flow programming.

11.5 TRACE ANALYSIS

Another area of great interest to the analyst is that of trace analysis, where small amounts of the compound of interest must be analysed in the presence of a large amount of interfering matrix. Problems of this type have already been touched upon in the determination of the intermediates present as impurities in tartrazine. Other typical trace analysis problems that can be dealt with by HPLC include the determination of drug metabolites in body fluids and the analysis of pesticide residues in environmental samples. A comprehensive review of trace analysis has been published (63). Two techniques that may be of interest in trace analysis of mixtures of wide-ranging polarities are column switching and column back-flushing. In the former, a relatively large volume of sample is injected onto a short analytical column A (figure 11.11). The early-eluting components of interest pass into the longer analytical column B. The valves are then switched so that these components are held at the top of column B while the later-eluting components from column A are analysed directly. The flow is then once again diverted through column B and the trace components can be analysed.

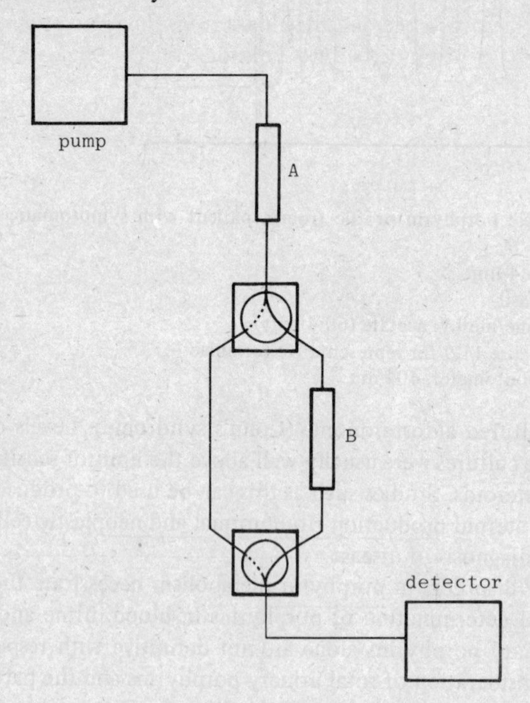

Figure 11.11. Schematic diagram of system for column switching.

Column back-flushing is used to analyse a trace component in the presence of a sample that is strongly retained in the chosen chromatographic system. After elution of the solute of interest the eluent flow is reversed via a series of valves (figure 11.12) and the strongly retained components stripped from the top of the column.

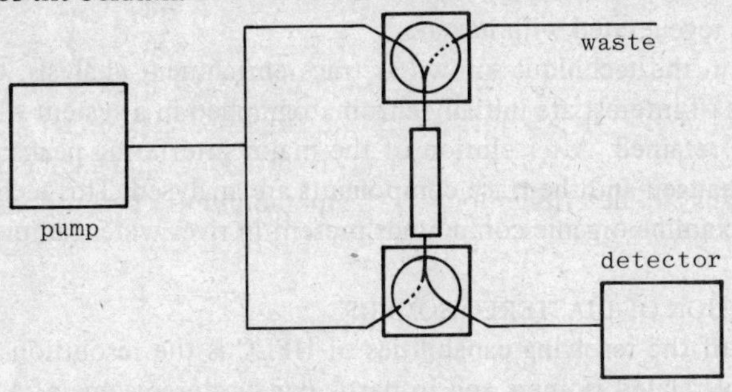

Figure 11.12. Schematic diagram of system for column back-flushing.

A combination of these two techniques has been used to analyse pesticide residues in milk (45). The milk extract is directly injected onto a 50 mm column of Partisil 5, and during T_1 (figure 11.13) the early-eluting pesticides HCB, DDE, DDT, TDE and α-BHC pass into the analytical column (150 mm column of Partisil 10). The valve is then switched and the eluate from the pre-column is led straight to the electron capture detector. The more polar pesticides,

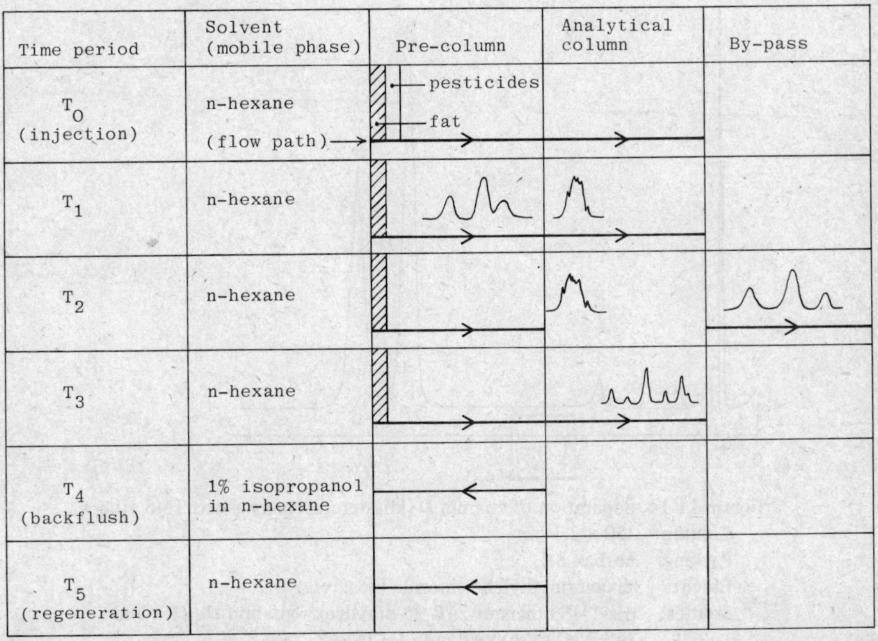

Time period	Solvent (mobile phase)	Pre-column	Analytical column	By-pass
T_0 (injection)	n-hexane (flow path)	pesticides fat		
T_1	n-hexane			
T_2	n-hexane			
T_3	n-hexane			
T_4 (backflush)	1% isopropanol in n-hexane			
T_5 (regeneration)	n-hexane			

Figure 11.13. Schematic representation of switching sequence used for analysis of pesticides in milk (64).

γ-BHC, β-BHC, heptachlorepoxide and dieldrin are analysed and then the flow is again switched through the analytical column. Once the first group of pesticides has been separated and analysed the primary pump is stopped and 1% isopropanol in hexane is back-flushed through the pre-column to remove the adsorbed fat, which is determined with a refractive index detector. Finally, the pre-column is regenerated with hexane.

Alternatively, in the technique known as trace enrichment analysis, the trace components of interest are initially chromatographed in a system where they are strongly retained. After elution of the major interfering peaks the mobile phase is changed and the trace components are analysed. This technique has been used to examine organic compounds present in river water samples (64).

11.6 RESOLUTION OF DIASTEREOISOMERS

An exacting test of the resolving capabilities of HPLC is the resolution of a mixture of closely related isomers and in particular diastereoisomers. A particularly interesting study was undertaken by Ikekawa and Koizumi on the separation of Vitamin D_3 metabolites and their analogues (65). All of the metabolites and their analogues could be separated by adsorption chromatography using elution with a gradient of methanol in methylene chloride. It was found that the trimethylsilylether (TMS) derivatives of the 24-epimers of 24,25-dihydroxy-vitamin D_3 could be separated using 2% methylene chloride in hexane (figure 11.14). Using this method the configuration of the 24-hydroxyl

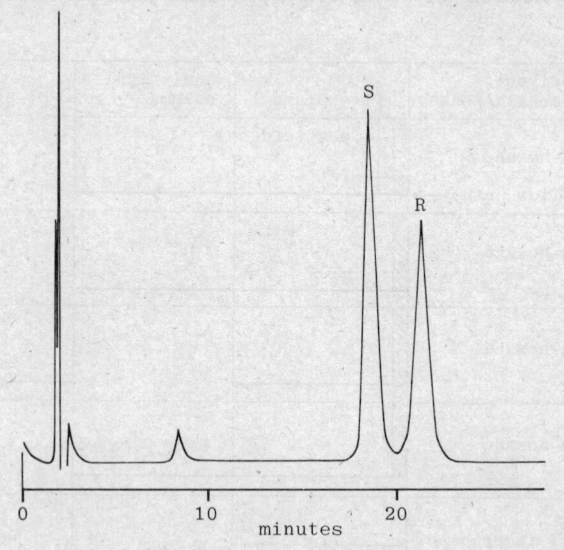

Figure 11.14. Separation of vitamin D_3 diastereoisomers as tris-TMS ethers.

Column: 250 x 2.1 mm
Packing: Zorbax SIL
Eluent: hexane/methylene chloride (98:2 v/v)
Solutes: tris-TMS ethers of 24R,25-dihydroxy-vitamin D_3 (R), and 24S,25-dihyroxy-vitamin D_3 (S)
Pressure: 93 bar
Detector: UV photometer, 254 nm

group of natural 24,25-dihydroxy-vitamin D_3 was determined to be R, using co-chromatography with the ^3H-labelled compound obtained biologically. The epimers of 24-hydroxy-vitamin D_3 can also be separated as their TMS derivatives, but the TMS-derivatives of the 24-epimers of 1*α*,24-dihydroxy-vitamin D_3 and 1*α*,24,25-trihydroxy-vitamin D_3 were unresolved. However, those epimers could be separated as their free forms. In this case the 24R isomer was eluted before the 24S isomer, but with the TMS derivatives the reverse elution order was observed.

These are only a few examples of the wide variety of applications that can be achieved by HPLC. It is hoped that they give an indication of the versatility of the technique.

REFERENCES

1. Wragg, J.S. and Johnson, G.W., *Pharmaceutical Journal 213* (1974) 601.
2. Bailey, F., *J. Chromatogr. 122* (1976) 73.
3. Wheals, B.B. and Jane, I., *Analyst 102* (1977) 625.
4. Dixon, P.F., Stoll, M.S. and Lim, C.K., *Ann. Clin. Biochem. 13* (1976) 409.
5. Wheals, B.B., *J. Chromatogr. 122* (1976) 85.
6. Brown, P.R., in *Advances in Chromatography*, vol. 13 (ed. Giddings, J.C.). Marcel Dekker, New York, 1975.
7. Brown, P.R., *High Pressure Liquid Chromatography. Biochemical and Biomedical Applications.* Academic Press, New York, 1973.
8. Done, J.N., Knox, J.H. and Loheac, J., *Applications of High Speed Liquid Chromatography.* John Wiley, London, 1974.
9. Rajcsanyi, P.M. and Rajcsanyi, E., *High Speed Liquid Chromatography.* Marcel Dekker, New York, 1975.
10. Pryde, A. and Gilbert, M.T., *Applications of High Performance Liquid Chromatography.* Chapman and Hall, London, 1978.
11. Ripphahn, J. and Halpaap, H., *J. Chromatogr. 112* (1975) 81.
12. Bailey, F. and Brittain, P.N., *J. Chromatogr. 83* (1973) 431.
13. Henry, R.A. and Schmit, J.A., *Chromatographia 3* (1970) 116.
14. Olson, M.C., *J. Pharm. Sci. 62* (1973) 2001.
15. Brown, N.D., Lofberg, R.T. and Gibson, T.P., *J. Chromatogr. 99* (1974) 635.
16. Weinberger, M. and Chidsey, C., *Clin. Chem. 21* (1975) 834.
17. Chen, C-T, Hayakawa, K., Imanari, T. and Tamura, Z., *Chem. Pharm. Bull 23* (1975) 2173.
18. Knox, J.H. and Jurand, J., *J. Chromatogr. 142* (1977) 651.
19. Lawrence, J.F. and Frei R.W., *Chemical Derivatisation in Liquid Chromatography.* Elsevier, Amsterdam, 1976.
20. Fitzpatrick, F.A., Siggia, S. and Dingman, J., *Anal. Chem. 44* (1972) 2211.
21. Henry, R.A., Schmit, J.A. and Dieckman, J.F., *J. Chromatogr. Sci. 9* (1971) 513.
22. Fitzpatrick, F. and Siggia, S., *Anal. Chem. 45* (1973) 2310.
23. Borch, R.F., *Anal. Chem. 47* (1975) 2437.
24. Durst, H.D., Milano, M., Kikta, E.J., Connelly, S.A. and Grushka, E., *Anal Chem. 47* (1975) 1797.
25. Pei, P.T.S., Kossa, W.C., Ramachandran, S. and Henly, R.S., *Lipids 11* (1976) 814.
26. Cooper, M.J. and Anders R.W., *Anal. Chem. 46* (1974) 1849.
27. Morozowich, W. and Douglas, S.L., *Prostaglandins 10* (1975) 19.
28. Fitzpatrick, F.A., *Anal. Chem. 48* (1976) 499.
29. Zimmerman, C.L., Pisano, J.J. and Appela, E., *Biochem. Biophys. Res. Commun. 55* (1973) 1220.
30. Graffeo, A.P., Haag, A. and Karger, B.L., *Anal. Lett. 6* (1973) 505.
31. Frank, G. and Strubert, W., *Chromatographia 6* (1973) 522.
32. Frankhauser, P., Fries, P., Stahala, P. and Brenner, M., *Helv. Chim. Acta. 57* (1974) 271.
33. Haag, A. and Langer, K., *Chromatographia 7* (1974) 659.
34. de Vries, J.Y., Frank, R. and Birr, C., *Febs. Lett. 55* (1975) 65.

35. Matthews, E.W., Byfield, P.G.H. and MacIntyre, I., *J. Chromatogr. 110* (1975) 369.
36. Bollet, C. and Caude, M., *J. Chromatogr. 121* (1976) 323.
37. Lawrence, J.F. and Frei, R.W., *J. Chromatogr. 98* (1974) 253.
38. Frei, R.W., Santi, W. and Thomas, M., *J. Chromatogr. 116* (1976) 365.
39. Muusze, R.G. and Huber, J.F.K., *J. Chromatogr. Sci. 12* (1974) 779.
40. Frei, R.W., Michel, L. and Santi, W., *J. Chromatogr. 126* (1976) 665.
41. Koen, J.G., Huber, J.F.K., Poppe, H. and den Boef, G., *J. Chromatogr. Sci. 8* (1970) 192.
42. Kissinger, P.T., Refshauge, C., Dreiling, R. and Adams, R.N., *Anal. Lett. 6* (1973) 465.
43. Fleet, B. and Little, C.J., *J. Chromatogr. Sci. 12* (1974) 747.
44. Willmot, F.W. and Dolphin, R.J., *J. Chromatogr. Sci. 12* (1974) 695.
45. Dolphin, R.J., Willmot, F.W., Mills, A.D. and Hoogeveen, L.P.J., *J. Chromatogr. 122* (1976) 259.
46. Scott, R.P.W., Scott, C.G., Munroe, M. and Hess, J., *J. Chromatogr. 99* (1974) 396.
47. Arpino, P.J., Dawkins B.G. and McLafferty, F.W., *J. Chromatogr. Sci. 12* (1974) 574.
48. Jones, P.R. and Yang, S.K., *Anal. Chem. 47* (1975) 1000.
49. McLafferty, F.W., Knutti, R., Venkataraghavan, R., Arpino, P.J. and Dawkins, B.G., *Anal. Chem. 47* (1975) 1503.
50. Carroll, D.I., Dzidic, I., Stillwell, R.N., Haegele, K.D. and Horning, E.C., *Anal. Chem. 47* (1975) 2369.
51. McFadden, W.H., Schwartz, H.L. and Evans, S., *J. Chromatogr. 122* (1976) 389.
52. Roos, R.W., *J. Pharm. Sci. 61* (1972) 1979.
53. Bakalyar, S.R. and Henry, R.A., *J. Chromatogr. 126* (1976) 327.
54. Rosenbaum, D., *Anal. Chem. 46* (1974) 2226.
55. Wittmer, D.P., Nuessle, N.O. and Haney, W.G., *Anal. Chem. 47* (1975) 1422.
56. Knox, J.H. and Laird, G.R., *J. Chromatogr. 122* (1976) 17.
57. Pound, N-J. and Sears, R.W., *J. Pharm. Sci. 64* (1975) 284.
58. Midha, K.K., McGilveray, I.J. and Charette, C., *J. Pharm. Sci. 63* (1974) 1234.
59. Knox, J.H. and Jurand, J., *J. Chromatogr.*, in Press.
60. O'Hare, M.J., Nice, E.C., Magee-Brown, R. and Bullman, H., *J. Chromatogr. 125* (1976) 357.
61. Evans, N., Jackson, A.H., Matlin, S.A. and Towill, R., *J. Chromatogr. 125* (1976) 345.
62. Gray, G.H., Lim, C.K. and Nicholson, D.C., in *HPLC in Clinical Chemistry* (eds. Dixon, P.F., Gray, C.H., Lim, C.K. and Stoll, M.S.) p. 79. Academic Press, London, 1976.
63. Kirkland, J.J., *Analyst 99* (1974) 859.
64. Little, J.N. and Fallick, G.J., *J. Chromatogr. 112* (1975) 389.
65. Ikekawa, N. and Koizumi, N., *J. Chromatogr. 119* (1976) 227.

12. COLUMN PACKING AND TESTING

To achieve high kinetic performance in HPLC it is necessary to be able to pack columns reproducibly, to have a yardstick of performance and to have a standard test procedure for determining the performance.

12.1 COLUMN PACKING: PRINCIPLES

Modern HPLC columns are packed with microparticles whose mean diameters are in the range 3 to 20 μm, the commonest mean sizes being 5 and 10 μm. Commercial batches of HPLC materials will normally contain 95% of their particles within a two- or three-fold diameter range centred on the nominal size. A few materials have closer distributions. The most important feature of the size distribution curve for a good material is a sharp cut-off at the low end of the range and the absence of fines.

In the past, particles down to about 30 μm have been successfully dry-packed using the rotate, bounce and tap method (1,2). This technique is not generally suitable for particles below 20 μm diameter, where a slurry technique is normally used (3–7). These techniques have recently been reviewed by Webber and McKerrell (6) and by Bristow *et al*. (7).

There are four main requirements for a satisfactory slurry-packing procedure: (a) particles must not agglomerate; (b) particles must not sediment too fast during the procedure; (c) particles must hit the accumulating bed at a high impact velocity; and (d) the bed must be packed under high compression. There are undoubtedly many ways in which these requirements can be met, and there is no procedure that is clearly superior to others. Thus each chromatographer is likely to have his own slurry-packing method and will tend to proclaim its special virtues.

Some general points can, however, be made. Initial dispersion of the individual particles of packing and prevention of subsequent agglomeration is vital. Small particles are generally charged and so will naturally repel each other once separated. Silica itself is negatively charged at pH >5 and positively charged at pH <5. It is always packed under conditions where it is negatively charged. Thus after initial dispersion, which is best carried out in an ultrasonic bath, slurries should be stable. Silica is especially well dispersed in methanol, which wets the surface well. If a water-immiscible slurry liquid is used some methanol may have to be added to wet the silica. Reversed-phase materials can be dispersed either in hydrophobic liquids, e.g. hexane, carbon tetrachloride, etc., or in methanol or even methanol/water mixtures. If a new procedure is being developed it is important to check the stability of the dispersion. Firstly the suspension should be allowed to stand for a few minutes. After this time on gentle agitation there should be no evidence of "curdling". Secondly the suspension should be examined under a microscope, to ensure that particles do not stick together on colliding.

Whereas charge on particles is necessary for the stability of the dispersion, excessive charge can produce problems, especially with reversed-phase bonded materials that have negligible electrical conductivity. During packing of such materials with a hydrophilic liquid an electrokinetic charge develops due to the flow over the particles. This results in the particles acquiring a very high potential. They therefore repel each other, and this repulsion may be so great that an unstable bed results which collapses once the flow is stopped. This is a major cause of irreproducibility and poor packing of bonded materials. The problem is simply overcome by incorporating a low concentration of an electrolyte into the slurrying liquid; this enables the electrokinetic charge to leak away. A concentration of around 0.1 to 1 gm dm^{-3} of, say, sodium acetate in the slurrying liquid is sufficient.

Table 12.1. Sedimentation velocities of silica gel particles.

		Hexane	Water	Glycol
η ($N\ s\ m^{-2}$)		0.33×10^{-3}	1.0×10^{-3}	20×10^{-3}
$\Delta\rho$ ($kg\ m^{-3}$)		0.77×10^{3}	0.60×10^{3}	0.54×10^{3}
u ($mm\ min^{-1}$)	5 μm	2	0.5	0.02
	10 μm	8	2	0.09
	20 μm	31	8	0.36

Sedimentation of particles during the packing procedure must be minimised, as this tends to produce size fractionation across the column. The rate of sedimentation, u, for spherical particles is given by the Stokes equation

$$u = \frac{1}{18} \frac{d_p^2(\rho_s - \rho_l)g}{\eta} \tag{12.1}$$

where ρ_s is the density of the liquid filled particles, ρ_l the density of the supporting liquid, g the gravitational acceleration and η the liquid viscosity. Typical sedimentation rates for particles of different sizes in hexane, water and ethylene glycol are given in table 12.1

The effects of sedimentation are minimised by carrying out the procedure in the shortest time, by using small particles (say 5 μm rather than 10 μm particles), by using a liquid of high viscosity and by using a liquid of density close to that of the solid part of the particles (that is 2.2 g cm^{-3} for silica gel). Slurries in which the density of the liquid is the same as that of the solid are called balanced-density slurries. The composition of some supporting liquids with a density around 2.2 are listed in table 12.2. They were widely used initially (3,4) but are now less popular due to high cost and toxicity.

On the basis of criterion (b) one would advocate the use of a liquid of fairly high density and viscosity, although it is worth noting that when using a constant driving pressure the decreased flow rate with a viscous liquid would exactly cancel out the decreased sedimentation rate. Thus no advantage would accrue from increased viscosity except over the period of preparation before the slurry was actually driven into the column.

In order to obtain a stable packed column it is important that the particles hit the bed at an adequate velocity and that the viscous force pressing them onto the bed and compressing the bed itself is high (criteria c and d). High packing velocity will be achieved by using high packing pressures or flow rates in conjunction with non-viscous solvents. High viscous force will be achieved by high packing pressure or flow rate, irrespective of viscosity. If very viscous solvents are used the impact velocity may be low unless extreme pressures are used. Nevertheless such supporting liquids have been successfully used by Asshauer and Halasz (5), although the great majority of chromatographers use supporting liquids of only moderate viscosity (say 0.3 to 2×10^{-3} N s m^{-2}).

Table 12.2. Component liquids for balanced-density slurries (densities in g cm^{-3}).

Dense liquid		Additive liquids				% composition to give $\rho = 2.2$		
A	ρ	B	ρ	C	ρ	A	B	C
$C_2H_2Br_4$	2.97	C_2Cl_4	1.62	CH_3OH	0.80	44	56	0
						49	41	10
$CHBr_3$	2.89	C_2Cl_4	1.62	CH_3OH	0.80	52	38	10
CH_3I	2.28	CH_3OH	0.80		–	95	5	
$C_2H_4Br_2$	2.18		–		–	100		

Approx. relative costs: CH_3OH 1.0; C_2Cl_4 1.2; $C_2H_4Br_2$ 2.3
$C_2H_2Br_4$ 20; $CHBr_3$ 22.5; CH_3I 25.

Discussion as to whether it is best to pack columns using a constant-pressure pump or a constant-flow pump have been inconclusive, and equally good columns can be packed with either type of pump if the flow rate is kept above 5 or 10 cm^3 min^{-1} throughout the procedure and if the final pressure is at least 100 bar per 100 mm length of column. The final pressure (or flow rate) during packing should, of course, exceed the maximum operating pressure (or flow rate) of the column. The maximum usable pressure is set by the equipment or by the break-up of the packing (somewhere about 1000 bar, normally). There is likewise no clear agreement as to the best ratio of liquid to solid that should be used, although few chromatographers employ liquid to solid ratios greater than 50:1 or less than 5:1. When the ratio is small there is less time for sedimentation but there is possibly a greater tendency for clumping. Stirring the slurry is advisable if long columns are to be packed with dilute slurries.

There is a wide variety in the geometry of packing chambers used to hold the slurry and again the precise configuration does not seem too important provided that there is a reasonably smooth lead-in to the column. Some variants, which all give good results, are shown in figure 12.1.

According to Bristow (7), during upward packing of a column the surface

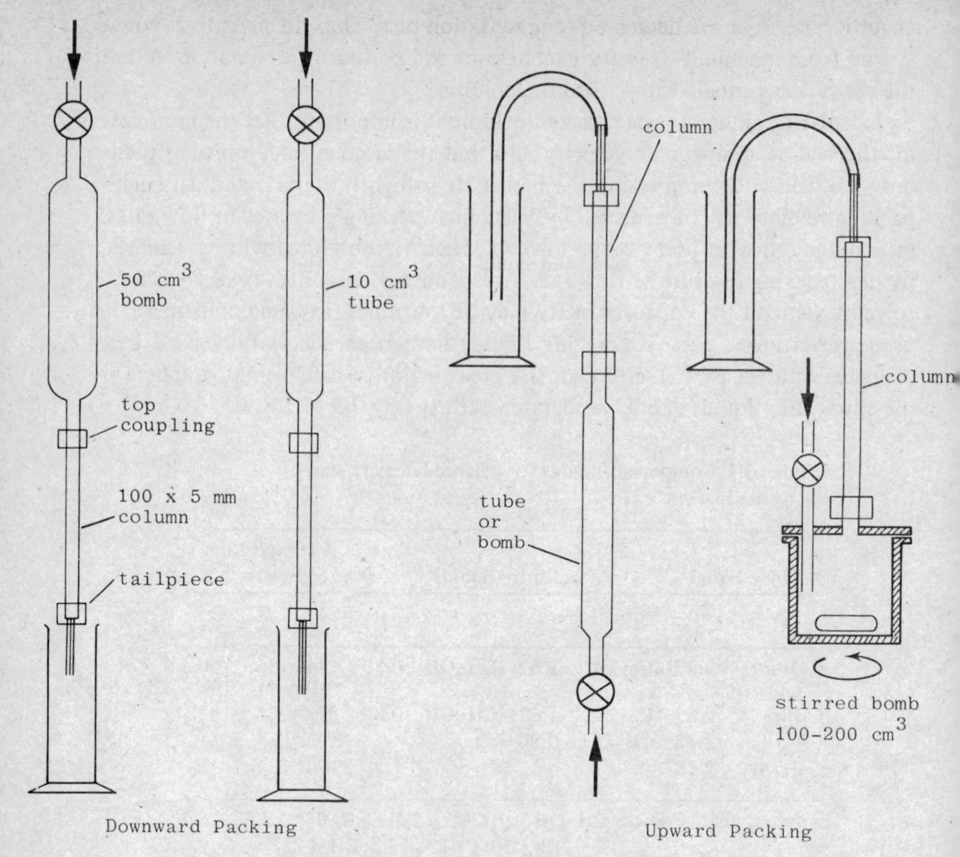

Figure 12.1. Illustrative arrangements for slurry packing of HPLC columns.

of the bed remains strictly horizontal while during downward packing it becomes domed as the flow velocity falls. In our own laboratories we have observed that upward packing gives consistently more efficient and reproducible columns when valve injection is employed, although with syringe injection all methods appear equivalent. We interpret this to mean that upward packing is less subject to the effects of sedimentation and trans-column fractionation so that the packing is more uniform across the column section. With valve injection the sample is distributed over the entire cross-section of the column, so trans-column uniformity is vital: with syringe injection it is only the central core that must be uniformly packed.

The chemical nature of the supporting liquid may be important. Thus balanced-density liquids (see table 12.2) normally contain brominated or iodinated alkanes. These will react with certain bonded phases, for example amino phases, and may be difficult to elute from others. They are also toxic. When packing silica gel using methanol, or any alcohol, the adsorbed alcohol may be difficult to remove. A column packed using methanol may appear to have different adsorptive characteristics from one packed with a non-hydrogen-bonding liquid. A final consideration in choosing the slurry liquid arises if the packing equipment is to be used to pack both silica gels and reversed-phase materials. The same supporting liquid is then advisable for both types of material.

Table 12.3 gives a number of slurry liquids that have been shown to provide high efficiency columns with different materials packed into 100 × 5 mm columns at a pressure of 200 bar.

Table 12.3 Effective support liquids for slurry packing

Packing material	Supporting liquid	Slurry[a]	Ref.
Silica gel	methyl iodide/ methanol 90:10	C	3[b]
	methanol	C or D	7[b]
	0.001 M aqueous ammonia	D	8
Bonded silicas	hexane	C	9
(hydrocarbon,	carbon tetrachloride	C or D	6
amino, cyano)	methanol + 0.2 g dm^{-3} sodium acetate	C or D	b
	methanol/water 80:20 + 0.2 g dm^{-3} sodium acetate	C or D	b
Bonded ion exchangers	acetone	C	10

[a] Concentrated slurries (C) have a liquid:solid ratio of about 5, dilute slurries (D) a ratio of about 50.

[b] Methods found to be effective in Wolfson LC Unit. We do not have direct experience of the other methods but there is no reason to doubt that they are equally effective.

12.2 PACKING PROCEDURE

In this section we describe in some detail specific packing procedures that are used in the authors' laboratory and have been shown to give consistently reproducible results over a large number of columns. The procedures are certainly not unique and, as has already been noted, other procedures are likely to be equally effective.

The packing equipment that has been used most widely is shown in figure 12.2. The pump is a large-volume air-driven intensifier pump capable of generating 200 bar, with a nitrogen pressure of about 10 bar. The details of the packing chamber and standard column are shown in figure 12.3.

For a 100 × 5 mm column approximately 2 g of packing material is dispersed in 10 cm^3 of supporting liquid. A satisfactory universal supporting liquid for silica gel and bonded silica gels is methanol, which for bonded phases should contain 0.2 g dm^{-3} of sodium acetate. The mixture of packing and supporting liquid is shaken vigorously in a small sample bottle and then placed for 5–10 min in an ultrasonic bath to effect complete dispersal of the packing. Meanwhile the column with its bottom gauze and tail fitting in place is filled with the support liquid. The packing chamber is attached and the whole mounted vertically with the column outlet downwards. The slurry is removed from the ultrasonic bath,

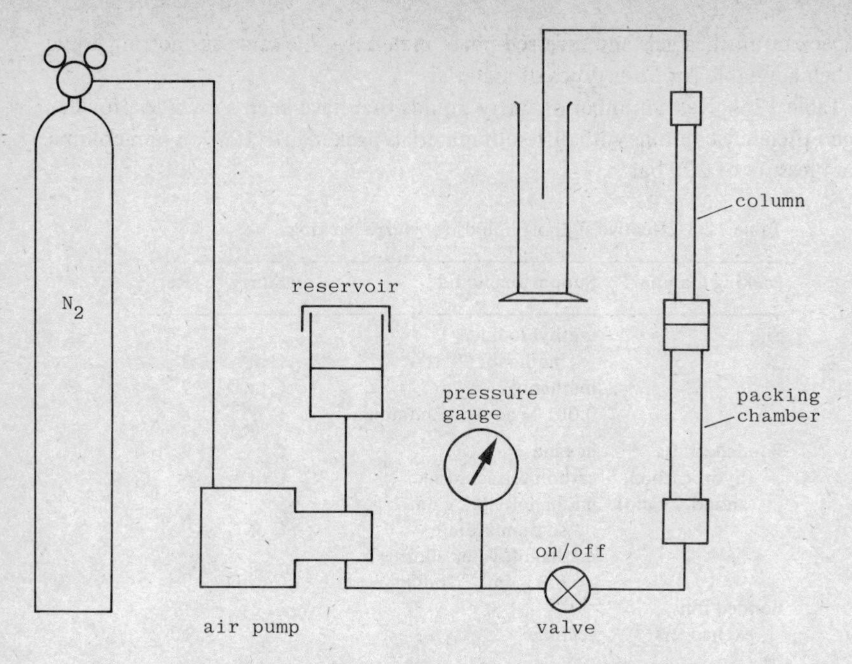

Figure 12.2. Slurry packing system using a pneumatic intensifier pump.

shaken vigorously and poured into the packing chamber. The chamber is topped up, if necessary, with support liquid and the lead from the pump connected. The pump is pressurised up to the on/off valve to a liquid pressure of 200 bar or more (see figure 12.1).

The column and packing chamber are now inverted and the column firmly clamped. The valve is opened and 100–200 cm^3 of the support liquid or a suitable follower liquid is pumped through the column at the maximum pressure. Clear liquid should, of course, emerge from the column. With 5 μm material the flow rate at 200 bar should finally be around 15 cm^3 min^{-1} if pure methanol is used as support or driver liquid. A much lower flow rate than this indicates a blockage of either the outlet tube or the gauze terminating the column. Fines in the packing material may be suspected.

When the desired amount of liquid has been passed the high pressure valve is closed and the column inverted again. The air pump is depressurised, the liquid pressure is allowed to fall to zero and the column left for 5 minutes before carefully disconnecting it from the packing chamber.

The top of the column packing is now made level with the shelf by carefully removing excess packing material with a micro spatula. The top of the packing is finally smoothed with a PTFE tamping rod. The top gauze, PTFE ring and glass beads are put in place. The column is then ready for use after conditioning with the desired eluent.

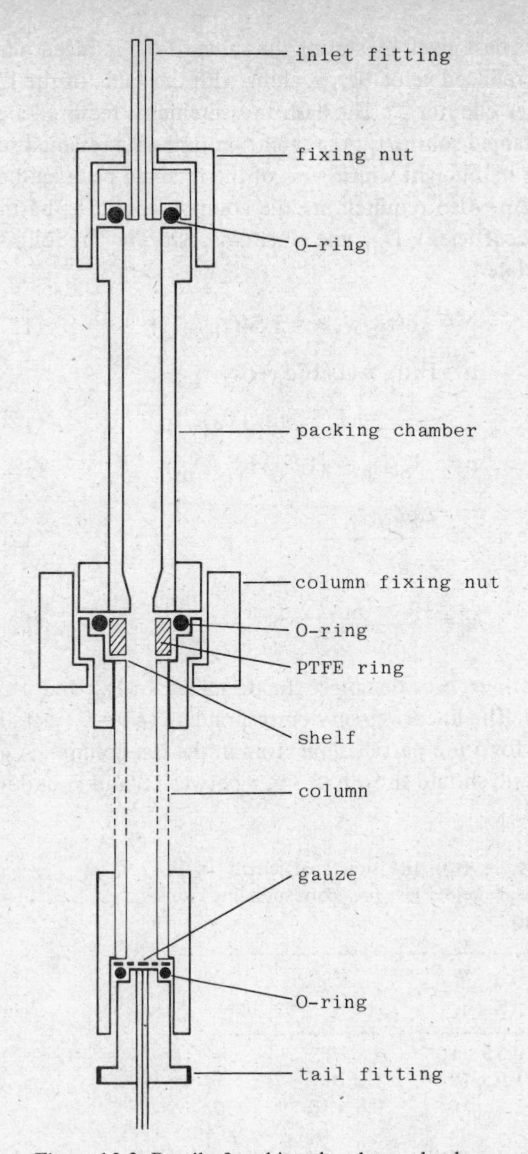

Figure 12.3. Detail of packing chamber and column.

12.3 COLUMN TESTING

It cannot be emphasised too strongly that before a column is put into use in a research or routine laboratory situation its efficiency should be determined by a standard procedure. Only in this way will it be possible to monitor deterioration in performance at a later stage or to assess whether a poor separation obtained, say, in a new application, is due to a poorly packed column or to the use of the wrong stationary phase or eluent. For the purpose of column testing one should use a simple test mixture designed to show the column performance at its best (11).

Column performance is best given by stating the value of the reduced plate height, h, at a number of reduced velocities, ν, along with the value of the flow resistance parameter ϕ' (see chapter 2). The basic measurements required are the retention time of an unretained solute, t_o; the retention time of a retained solute, t_R; the base width, w_t, or half height width, $w_{1/2}$, of the retained peak; and the operating pressure drop Δp. Also required are the column length, L; particle size, d_p; solute diffusion coefficient, D_m; and eluent viscosity, η. The following parameters are then calculated:

Plate number $$N = 16(t_R/w_t)^2 = 5.54(t_R/w_{1/2})^2 \quad (12.2)$$

Reduced plate height $$h = H/d_p = (L/16d_p)(w_t/t_R)^2 = (L/5.54d_p)(w_{1/2}/t_R)^2 \quad (12.3)$$

Reduced velocity $$\nu = ud_p/D_m = (L/t_o)(d_p/D_m) \quad (12.4)$$

Column flow resistance parameter $$\phi = \frac{\Delta p d_p^2 t_o}{\eta L^2} \quad (12.5)$$

Phase capacity ratio $$k' = \frac{t_R - t_m}{t_m} \quad (12.6)$$

To evaluate ν and ϕ' for test purposes the approximate values for D_m and η given in table 12.4 may be used. The linear velocity corresponding to $\nu = 7$ (just above that for minimum h) and for 5 μm particles are given in the last column. A good column, as previously noted, should show $h \approx 3$ at ν between 3 and 7, and have ϕ' in the range 500–1000.

Table 12.4 Viscosities (η), diffusion coefficients (D_m) and optimum linear velocities (u, corresponding to $\nu = 7$ for $d_p = 5$ μm)

Eluent	η (N s m^{-2})	D_m (m^2 s^{-1})	u (mm s^{-1})
Hexane	0.33×10^{-3}	4×10^{-9}	4
Water	1.0×10^{-3}	1×10^{-9}	1
Methanol/water (40:60 v/v)	1.8×10^{-3}	0.6×10^{-9}	0.6
Methanol	0.6×10^{-3}	1.9×10^{-9}	1.5

In general, diffusion coefficients in m^2 s^{-1} may be estimated with sufficient accuracy for calculation of reduced velocities by the Wilke-Chang equation (11, 12):

$$D_m = \frac{7.4 \times 10^{-15} (\psi M)^{1/2} T}{V_s^{0.6} \eta} \quad (12.7)$$

M is the relative molecular weight of the eluent; ψ is an association constant whose value is 2.6 for water, 1.9 for methanol, 1.5 for ethanol, and 1.0 for unassociated solvents; T is the absolute temperature in K; V_s is the molar volume of the solute in $cm^3 mol^{-1}$; and η is the eluent viscosity in $N s m^{-2}$ (see appendix for values).

The performance of a column packed with a reversed-phase bonded silica is conveniently evaluated using the following operating conditions:

Eluent:	methanol/water, 40:60 v/v (short-chain bonded hydrocarbon) or 60:40 v/v (long-chain bonded hydrocarbon)
Test mixture:	acetone, phenol, p-cresol, 3,5-xylenol, anisole, phenetole (dissolved in eluent containing excess methanol)
Sample volume:	1–5 mm^3
Injection:	syringe or valve
Temperature:	ambient
Pressure:	13–20 bar (200–300 psi) for a 100 mm column
Detector:	UV photometer set at λ = 254 nm and 0.1 AUFS
Chart speed:	30 mm min^{-1}

For columns packed with adsorbents the following eluent and test mixture can conveniently be used, other operating conditions remain the same:

Eluent:	hexane/methanol 99.5:0.5 v/v
Test mixture:	pentane, toluene, nitrobenzene, acetophenone, 2,6-dinitrotoluene, 1,3,5-trinitrobenzene (dissolved in eluent)

The sample is injected either by syringe or injection valve. An injection mark is made on the recorder trace by quickly switching the photometer or recorder sensitivity knob. The chart speed must be fairly high so that accurate measurement of peak width can be made. The following data should then be recorded on the test sheet:

1. Column bed length, L in mm (to nearest mm)
2. Column bore, d_c in mm (to nearest 0.1 mm)
3. Nominal particle diameter, d_p, in μm
4. Pressure drop Δp
5. Chart speed in mm min^{-1}
6. Retention distance or time for an unretained solute (usually a refractive index peak), t_m, in min or s (to nearest mm or 1%)
7. Retention distance or time for retained solutes, t_R, in min or s to each solute peak (to nearest mm or 1%)
8. Peak widths for retained solutes, w_t or $w_{1/2}$, in min or s (to nearest 0.1 mm)

Typically, a 100 × 5 mm column packed with 5 μm material should give between 5000 and 8000 plates under the above test conditions, corresponding to a reduced plate height of between 2.5 and 4. This applies to both adsorbents and bonded materials.

REFERENCES

1. Sie, S.T. and van den Hoed, N., *J. Chromatogr. Sci.* *7* (1969) 257.
2. Done, J.N., Knox, J.H. and Loheac, J., *Applications of High Speed Liquid Chromatography*. John Wiley, London, 1974.
3. Kirkland, J.J., *J. Chromatogr. Sci.* *9* (1971) 206.
4. Majors, R.E., *Anal. Chem.* *44* (1972) 1722.
5. Asshauer, J. and Halasz, I., *J. Chromatogr. Sci.* *12* (1974) 139.

6. Webber, T.J.N. and McKerrell, E.H., *J. Chromatogr. 122* (1976) 243.
7. Bristow, P.A., Brittain, P.N., Riley, C.M. and Williamson, B.F., *J. Chromatogr. 131* (1977) 57.
8. Kirkland, J.J., *J. Chromatogr. Sci. 10* (1972) 593.
9. Herbert, N.C., private communication.
10. Cox, G.B., Loscombe, C.R., Slucutt, M.J., Sugden, K. and Upfield, J.A., *J. Chromatogr. 117* (1976) 269.
11. Bristow, P.A. and Knox, J.H., *Chromatographia 10* (1977) 279.
12. Wilke, C.R. and Chang, P., *Amer. Inst. Chem. Engr. J. 1* (1955) 264.

13. ILLUSTRATIVE EXPERIMENTS

The following series of experiments aims to provide a selection of practical procedures to illustrate topics and techniques discussed in earlier sections of the book. With one exception the experiments were developed by manufacturers and tested in the laboratory as part of intensive practical courses on HPLC given at Edinburgh in Summer 1977. Although by no means comprehensive, this collection of experiments covers several basic features of LC and, with minor changes, they may be applied to any convenient high-performance separation, not only those suggested in this chapter. As described, each experiment may be satisfactorily completed within an hour or so, and may to advantage be extended to explore the topic in greater depth. Where possible the experimental results that are reported were obtained using equipment provided by manufacturers at the Edinburgh HPLC courses. The authors wish to thank the manufacturers both for the loan of their equipment and for contributing many of the basic experimental ideas.

The range of chromatographic features illustrated by the experiments is intended to cover the more relevant problem areas and techniques in current LC practice. Thus experiments have been included on the effects of dead volume and injection technique on chromatographic performance. The effects of eluent flow rate and variable UV-spectrometer wavelength are also explored. Stop-flow spectral scanning and the use of a UV detector combined with a flow fluorimeter are included as examples of the sensitivity and selectivity offered by spectrometric detectors and their usefulness in characterising eluted solutes.Quantitative analytical procedures in LC are discussed and their precision is assessed. The reproducibility of the loop valve injector is examined as part of an exercise comparing a gradient elution method with isocratic conditions.

The range of applications covered by these experiments reflects the wide use of reversed-phase packing materials. The present selection includes aromatic insecticide intermediates (13.4.1), polynuclear aromatic compounds (13.3.2 and 3), phthalate plasticisers (13.1.2), a rodenticide (13.2.2) and several pharmaceutical applications: steroids (13.1.1), barbiturates (13.1.3), injections (13.2.1), anticonvulsants (13.3.1) and sulphonamides (13.4.2). The high proportion of methods for drugs reflects the expanding interest in this area, as indicated in chapter 11.

In the confines of this chapter it has not been possible to include experiments on several topics of relevance, e.g. temperature effects, reversal of elution order, non-spectroscopic detectors, or ion-pair and soap chromatography. However, it is hoped that those presented may serve as useful source material for training technical staff and perhaps for setting up demonstrations to impress visitors to the laboratory.

For convenience we have given relevant references at the end of each experiment rather than at the end of the chapter.

13.1 FACTORS AFFECTING CHROMATOGRAPHIC PERFORMANCE IN HPLC

13.1.1 *The Effect of Dead Volume on Column Efficiency*

As emphasised in chapter 8, an essential design feature for successful chromatography by HPLC is that dead volume, and particularly unswept dead volume, should be kept to a minimum by the use of suitably drilled out Swagelock fittings or specially machined connectors. A further requirement is that, for any connections between column and injector or detector, tubing of not more than 0.25 mm (0.010 in) bore is used. The lengths of tubing allowable for a maximum acceptable volume dispersion (30 mm^3) may be calculated by reference to Taylor's equation, as described in chapter 8. Thus, for 0.25 mm tubing the maximum acceptable length would be about 100 mm for typical flow conditions and solute diffusion in water.

This experiment illustrates the effect of dead volume on chromatographic performance for changes in the length and diameter of tubing to connect the column with the detector (A) and with the injector (B).

Equipment

Loop valve injector or septum injector (suitable for part A only)
Pump
Fixed or variable-wavelength UV detector with 10 mm^3 flow-cell
approx. 1 m lengths of microbore tubing, internal diameter 0.010 in (0.25 mm), 0.020 in (0.50 mm), and 0.040 in (1.0 mm)
Recorder

Column System

250 mm × 4.5 mm packed with 10 μm Spherisorb ODS

Mobile Phase

Acetonitrile/Water (85:15 v/v)

Sample

A mixture of testosterone esters in approximately equal concentrations (ca. 500 ng mm^{-3}) dissolved in mobile phase (elution order 1–3):

1. testosterone propionate
2. testosterone phenylpropionate
3. testosterone isocaproate

Typical Operating Conditions

Detector wavelength: 254 nm
Detector sensitivity: 0.3–0.5 AUFS
Pressure drop: 40 bar (600 psi)
Mobile phase flow rate: 3.0 cm^3 min^{-1}

Special Features

It is necessary to be able to replace the normal connection tubing with the additional tubing from the column to detector for part A and from injector to column for part B of the experiment. Depending on the type of equipment available, it may be helpful to use a "zero-dead-volume" coupling to connect the additional tubing to existing tubing by butting end to end.

Experimental Procedure

1. Set up the chromatographic system with degassed solvent and establish a stable base-line with the conditions suggested above.

2. Inject 10 mm^3 sample and adjust flow and detector sensitivity to give a chromatogram showing three well-resolved peaks for the esters.

3. Increase the recorder speed to approx. 5 cm min^{-1} and inject 10 mm^3 sample again to obtain a chromatogram showing large, easily measurable peaks.

4. After the chromatogram, turn off the solvent flow.

Part A: Dead volume between column and detector

5. Replace the short connection between column and detector with a 1 m length of 0.010 in tubing.

6. Inject 10 mm^3 sample and record the chromatogram on expanded scale as in 3 above.

7. Repeat steps 5 and 6, inserting the 1 metre lengths of 0.020 and 0.040 in tubing after turning off the solvent flow.

8. Finally, turn off solvent flow and re-connect the column directly to the detector using the normal short connection.

Part B: Dead volume between injector and column

9. Inject a 10 mm^3 sample, recording the chromatogram on expanded scale as in 3 above to check that the original column performance has been restored.

10. Turn off solvent flow and detach the injector from the column. Re-connect with 1 m of 0.040 in tubing.

11. Inject 10 mm^3 sample and record the chromatogram as in 3 above.

12. Stop solvent flow and re-connect the sample valve injector directly to the column as before.

Calculation of Results

1. Calculate the column efficiency in terms of the number of theoretical plates, N, as described in section 12.3, using the equation

$$N = 5.54 \left(\frac{t_R}{w_{1/2}}\right)^2 \qquad (1)$$

where t_R is the retention time (expressed as chart mm) and $w_{1/2}$ is the peak width at half-height (expressed as chart mm).

Taking the last peak of each chromatogram, measure t_R (to the nearest 0.5 mm), $w_{1/2}$ (to the nearest 0.2 mm) and calculate N for each set of experimental conditions, drawing the appropriate conclusions.

2. Calculate the volume dispersion (w_{tube}) for each tubing dimension, using Taylor's equation (see chapter 8 and ref. 1).

$$w_{tube} = 0.36\, d^2 \left(\frac{Lf_v}{D_m}\right)^{1/2} 10^9\ mm^3 \qquad (2)$$

where w_{tube} is the volume dispersion; f_v is the volume flow rate in $m^3\ s^{-1}$ [= 1/60 (flow rate in $cm^3\ min^{-1}$) $\times 10^{-6}\ m^3\ s^{-1}$]; D_m is the diffusion coefficient in $m^2\ s^{-1}$ for component in eluent (approximately $10^{-9}\ m^2\ s^{-1}$ for components in water); L is the length of tubing in m; and d is the internal diameter of tubing in m.

For example, with water as the eluent at a flow rate of 3.0 $cm^3\ min^{-1}$ (as suggested above), then for tubing 1 m long and 0.010 in (0.25 mm) internal diameter the volume dispersion attributable to this source would be calculated as follows:

$$f_v = \frac{3.0}{60}\ cm^3\ s^{-1} = 0.05 \times 10^{-6}\ m^3\ s^{-1} = 5 \times 10^{-8}\ m^3\ s^{-1}$$

$$D_m = 10^{-9}\ m^2\ s^{-1}$$

$$d = 0.010\ in = 0.25\ mm = 2.5 \times 10^{-4}\ m$$

Since 1 $m^3 = 10^9\ mm^3$

$$w_{tube} = 0.36\ (2.5 \times 10^{-4})^2 \left(\frac{1 \times 5 \times 10^{-8}}{10^{-9}}\right)^{1/2} \times 10^9 = 160\ mm^3$$

For a volume flow rate of 1.0 $cm^3\ min^{-1}$, the corresponding volume dispersion introduced by the same tube would be about 92 mm^3, which illustrates the profound role exercised by flow rate in determining peak dispersions,as discussed in experiment 13.1.2.

3. Measure the peak base intercept for peak 3 in each chromatogram by extrapolating tangents at the points of inflection to cut the baseline. Expressed as mm^3 eluent flow, this is an approximate measure of the total peak dispersion, w_{total}, for the component subjected to the conditions of increased dead volume.

As shown in chapter 8, the w_{total} is related to the sum of squares of each dispersive component in the system:

$$w^2_{total} = w^2_v + w^2_{app} + w^2_{tube} \qquad (3)$$

where w_v is the dispersion introduced by the column alone and w_{app} is the dispersion introduced by the equipment (connectors, detectors etc). Thus, the volume dispersion, w_{tube}, introduced by the added tubing would be related to the final peak dispersion consequently observed by the equation:

$$w^2_{total} = w^2_{total\ (initial)} + w^2_{tube} \qquad (4)$$

where $w_{total\ (initial)}$ is the measured peak dispersion for the system before adding extra dead-volume connectors.

A significant consequence of this relationship is that a volume dispersion increase of, say, 160 mm^3 would precipitate a drastic deterioration in column performance for a highly efficient column, while having little effect on a poor system. This is illustrated by the data in table 13.1, calculated for a component with V_R = 200 mm^3, eluted at 3.0 $cm^3\ min^{-1}$ with and without addition of 1.0 m $\times$ 0.25 mm tubing.

Table 13.1. Effect of dead volume (theoretical).

Initial plate number N	Initial w_{total} (mm^3)	Calculated w_{tube} (mm^3) (eqn 2)	Resultant w_{total} (mm^3) (eqn 4)	Increase in peak base width (%)	Final plate number N
10,000	80	160	180	125	2,000
2,500	160	160	225	41	1,260
1,000	250	160	300	20	710

4. Calculate the total peak dispersion anticipated for each added connector, taking the initial value of w_{total} as being the peak base intercept (in mm^3) for the column system before addition of dead volume connectors, and using equation 4 above.

5. Compare the experimental with the calculated values and comment.

Typical Results

The data illustrated in figure 13.1 were obtained using an Applied Chromatography Systems Model 750/03 double reciprocating pump with an ACS Model 750/10A fixed-wavelength UV detector. Chromatographic conditions were as described above.

The calculations for peak 3 eluted at 3.0 cm^3 min^{-1} (testosterone isocaproate) are summarised in table 13.2 for initial column conditions and for dead volumes introduced as 1.0 m × 0.5 mm and 1.0 m × 1.0 mm tubing between column and detector.

Table 13.2 Effect of dead volume (experimental).

Dead volume added	Observed w_{total} (mm^3)	Observed w_{tube} (mm^3) (eqn 4)	Calculated w_{tube} (mm^3) (eqn 2)	Observed N	Observed V_R (mm^3)
none	1950	–	–	1460	18,600
1000 x 0.5 mm = 200 mm^3	2100	780	640	1300	18,900
1000 x 1.0 mm = 800 mm^3	3000	2300	2560	655	19,200

Comments

1. The observed and caculated values of w_{tube} are in good agreement.

2. The additional retention volume is roughly equal to the volume of the added tubing.

3. The results quoted are for a relatively inefficient column system and, as anticipated, show that only massive added dead volume causes any significant loss of performance. Thus the relative magnitude of dead volume effects varies with initial column performance.

4. Similar results would be expected for dead volume introduced between injector and column. This can occur spontaneously as, for example, when packing

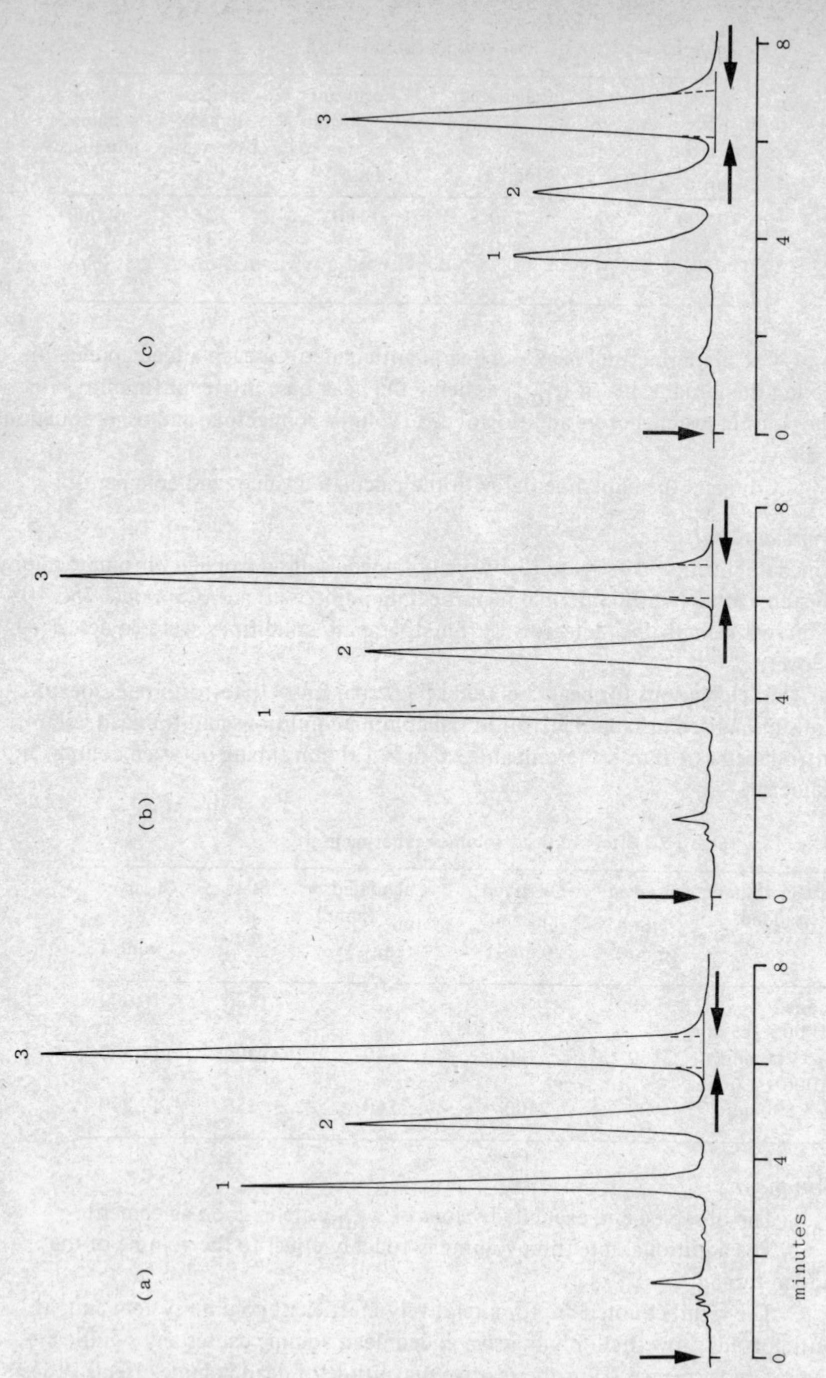

Figure 13.1. Effect of dead volume on resolution of testosterone esters on Spherisorb ODS. UV detection, 254 nm, 0.3 AUFS (full details in text). (a) normal connection between column and detector; (b) 1000 x 0.5 mm tube between column and detector; (c) 1000 x 1.0 mm tube between column and detector.

material at the top of a column settles to leave a cavity between injector and column.

5. A similar experiment could be devised to examine the relative merits of various types of micro-flow-cell, provided that an efficient separation was established initially.

Acknowledgement
This experiment is based on a method developed by Applied Chromatography Systems Ltd.

Reference
1. Taylor, Sir G., *Proc. Roy. Soc. A219* (1953) 186.

13.1.2 *The Effect of Eluent Flow Rate*

Eluent flow rate has a profound effect on column performance, as measured by the reducedplateheight h. The three main band dispersion processes (cf. chapter 2) contribute independently to plate height, which is related to eluent flow rate by a form of the van Deemter equation

$$h = \frac{H}{d_p} = \frac{L}{Nd_p} = A\nu^{\frac{1}{3}} + \frac{B}{\nu} + C\nu \qquad (1)$$

where d_p is the particle size in m; L is the column length in m; N is the number of theoretical plates; ν is the (dimensionless) reduced flow velocity (= ud_p/D_m, D_m being the diffusion coefficient in m^2s^{-1} and $u = L/t_o$ being the linear flow velocity in $m\ s^{-1}$); and A, B, C are constants.

A plot of log h versus log ν gives a characteristically shaped curve (cf. figure 2.4), which may be used to establish optimum flow conditions for a particular separation on a given column system.

Equipment
Pump
Loop valve injector (or septum injection system at lower pressures)
UV detector, fixed or variable-wavelength, with 10 mm^3 flow-cell
Recorder
Column System
250 × 6 mm packed with 10 μm LiChrosorb RP8 (C_8 bonded silica)
Mobile Phase
Acetonitrile/water (25:75 v/v)
Sample
A mixture of four components in methanol (listed in order of elution):

1. dimethylphthalate 1.5 $\mu g\ cm^{-3}$
2. diethylphthalate 1.5 $\mu g\ cm^{-3}$
3. biphenyl 0.1 $\mu g\ cm^{-3}$
4. terphenyl 0.3 $\mu g\ cm^{-3}$

Typical operating conditions
Detector wavelength: 254 nm
Detector sensitivity: 0.2 – 0.3 AUFS

Pressure drop: up to 170 bar (2500 psi)
Mobile phase flow rate: 1.0 to 4.0 $cm^3\ min^{-1}$

Special features

The flow rate should be accurately measured for each setting by using a stop-watch and volumetric flask to collect eluent.

Experimental procedure

1. Set up the chromatographic system with degassed solvent and establish a stable base-line at a flow rate of 1.0 $cm^3\ min^{-1}$. Adjust the UV detector sensitivity to give 80% full scale deflection for 10 mm^3 sample injection.

2. Increase recorder chart speed to 5 cm min^{-1} so that peaks are wide enough for accurate measurement.

3. Inject 10 mm^3 sample at an eluent flow rate of 1.0 $cm^3\ min^{-1}$, recording the chromatogram.

4. Increase flow rate to 2.0 $cm^3\ min^{-1}$, allow the base-line to stabilise and then inject 10 mm^3 sample as before.

5. Repeat 4 with a flow rate of 4.0 $cm^3\ min^{-1}$.

6. If time permits, the effect of a lower flow rate of 0.5 $cm^3\ min^{-1}$ should be examined. In this case, recorder speed should be reduced to about 2 cm min^{-1} (to conserve paper: the peaks will generally be broad enough to measure accurately).

Calculation of Results

1. Calculate the number of theoretical plates for the final peak in the chromatogram (terphenyl) at each flow rate (see chapter 12 for details), taking care to measure retention time (t_R) to the nearest 0.5 mm and peak width at half-height ($w_{1/2}$) to the nearest 0.2 mm.

2. Calculate the reduced plate height (h) for the particle size (d_p) and column length used (L).

3. The reduced linear velocity (ν) may be calculated by taking the diffusion coefficient for a small solute in water (table 12.4) $D_m = 1 \times 10^{-9}\ m^2\ s^{-1}$. Since the linear flow velocity $u = L/t_m$, where t_m is the unretained solute retention time in seconds, then

$$\nu = \frac{ud_p}{D_m} = \frac{L}{t_m} \frac{d_p}{D_m} = \frac{250 \times 10^{-3} \times 10 \times 10^{-6}}{t_m \times 10^{-9}} = \frac{2500}{t_m} \qquad (2)$$

t_m in seconds is usually readily measured from the chromatogram.

Plot log h versus log ν and compare with figure 2.4. Estimate the optimum eluent flow rate for the system.

Typical Results

The results for this experiment, performed on a Hewlett-Packard Model 1084A chromatograph (using the column system described above), are illustrated in figure 13.2 and summarised in table 13.3 for the final peak in each chromatogram.

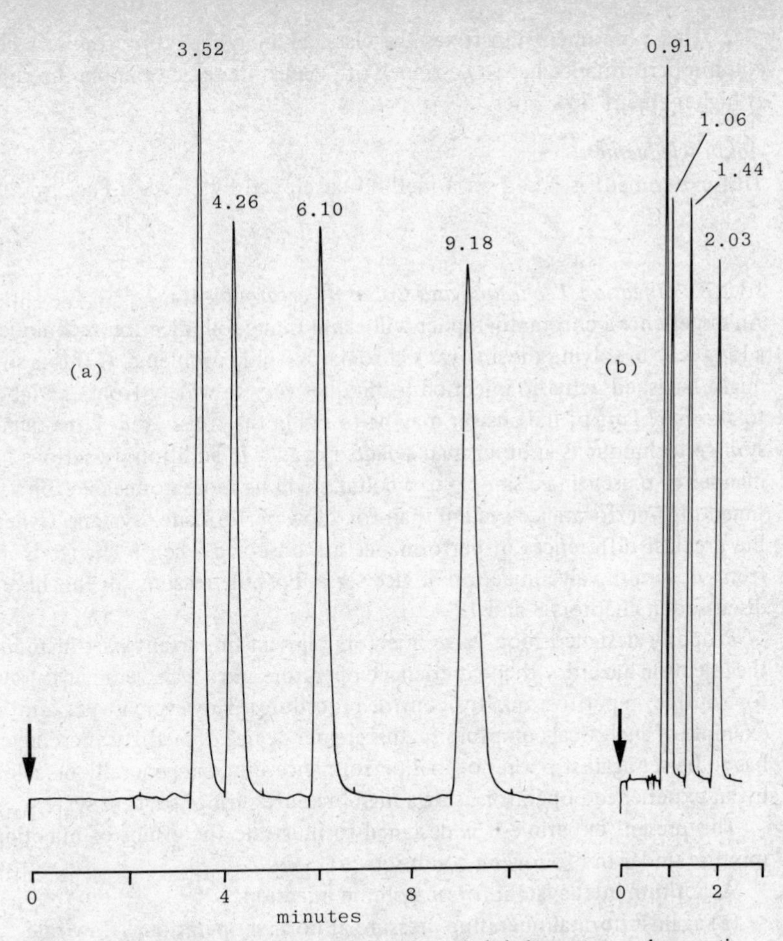

Figure 13.2. Effect of flow rate on resolution of phthalate esters and aromatic hydrocarbons on LiChrosorb RP8. UV detection, 254 nm, 0.26 AUFS (full details in text). Flow rates: (a) $f_V = 1.0\ cm^3\ min^{-1}$; (b) $f_V = 4.0\ cm^3\ min^{-1}$. Retention times are noted.

Table 13.3 Effect of flow rate.

f_V ($cm^3\ min^{-1}$)	t_m (s)	ν	t_R (min)	N	H (mm)	h
1.0	132	19	9.18	5020	0.050	5.0
2.0	71	35	4.12	4180	0.060	6.0
4.0	37	68	2.03	2280	0.110	11.0

Comments

1. Any HPLC system may be used for this experiment provided that a sample mixture is selected whose components elute within a reasonable time at the lowest eluent flow rate employed.

2. This experiment illustrates the classical analytical compromise where column performance has to be traded off against increased sample throughput at higher eluent flow rates.

Acknowledgement
This experiment is based on a method developed by Hewlett-Packard Ltd.

13.1.3 *Injection Technique and Column Performance*

An experienced chromatographer will claim that good injection technique goes a long way in solving the analyst's chromatographic problems. If this is so, it might be asked, why do injection techniques vary so widely from one laboratory to another? Part of the answer may be found in the truism that "one man's syringe technique is another man's band spread". In addition, variations in the manner of presenting a sample to a column will be more noticeable for a finely tuned high-performance system than for a low performance system. Generally, the greatest differences in performance are observed when a change is made from septum to valve injection or vice versa. Possible reasons for this have been discussed in chapters 8 and 12.

Properly designed loop valve injectors play an important part in reducing the injection hazard with inexperienced operators and are also eminently suitable for routine, repetitive quality control procedures. However, in yet another example of analytical compromise, this greater degree of security and convenience has to be set against poorer overall performance that can generally be achieved by an experienced operator using a high-pressure syringe-septum system.

The present experiment is designed to illustrate the effect of injecting a mixture under the following conditions:

A. Septum inlet system for on-column injection
(a) against normal operating pressure at normal operational flow rate
(b) stop-flow i.e. at zero flow rate
(c) at reduced flow rate
(d) as in (a) but delivering the sample over a 5-second interval
B. By loop valve injector at normal operational flow rate

Equipment
Pump
Loop valve injector
Low-pressure syringe
Septum injection system
High-pressure syringe
Fixed or variable-wavelength UV detector with 10 mm^3 flow-cell
Recorder
Column System
250 × 4.6 mm packed with 10 μm Partisil-ODS
Mobile Phase
Methanol/water (40:60 v/v)

Sample

Mixture of barbiturates (~ 1.5 μg cm^{-3}) in distilled water (listed in order of elution):

1. Barbituric acid
2. Sodium phenobarbitone
3. Calcium cyclobarbitone

Typical operating conditions

Detector wavelength: 254 nm
Detector sensitivity: 0.04 AUFS
Pressure drop: up to 100 bar (1500 psi)
Mobile phase flow rate: 0.5 to 2.0 cm^3 min^{-1}

Special features

A device to stop eluent flow must be fitted in the high-pressure eluent supply line to the column.

Although this experiment is designed for a valve loop volume of 3 mm^3 (corresponding to 3 mm^3 injected on column via the septum inlet), the volume of the valve loop may be higher, provided that whatever sample size is selected remains the same for septum and valve loop injection.

Experimental Procedure

1. Set up the chromatographic system with the septum injection system. Establish a stable base-line at the normal flow rate used (2.0 cm^3 min^{-1} for this system) and adjust detector sensitivity to give peaks between 60 and 80% full scale deflection.

2. Increase recorder chart speed to 5 cm min^{-1} to give adequate peak width for accurate measurement.

Part A: Septum injection system

3. Check that eluent flow rate is 2.0 cm^3 min^{-1}, inject 3 mm^3 sample (with the usual rapid, smooth plunger movement) and record the chromatogram.

4. Stop eluent flow, wait 30 s for pressure to subside, inject 3 mm^3 as before and re-start eluent flow immediately, marking the chromatogram at that point.

5. Reduce flow rate to half or less of the normal operating value and allow the base-line to stabilise. Displace the recorder pen from zero (to track subsequent base-line drift). Inject 3 mm^3 sample and immediately increase eluent flow rate to its normal value, marking the chromatogram at the point of injection.

6. Repeat 3 but control the rate of sample delivery so that the process of injection takes 5 s.

Part B: Loop valve injection system

7. Stop eluent flow and replace the septum injector with the loop valve injector and connector system.

8. Test the assembly for leaks and run a rapid test chromatogram.

9. With flow, recorder and detector conditions as for 3, introduce a 3 mm^3 sample and record the chromatogram.

Calculation of Results

1. Calculate the number of theoretical plates N for each set of conditions (see chapter 12) and compare.

2. Estimate any additional band dispersion introduced by the loop valve injector by reference to the equation for total band dispersion (w_{total}) in a chromatographic system (chapter 8):

$$w_{total}^2 = w_{initial}^2 + w_{lvi}^2$$

where $w_{initial}$ is the band dispersion observed for 3 above (calculated by measuring the peak base intercept and converting to mm^3 eluent) and w_{lvi} is the band dispersion attributable to the loop valve injector.

It should be noted that since the contribution of the septum injector to w_{total} is not taken into account in this calculation, this estimate of the effect of the loop valve injector is necessarily approximate.

Typical Results

Results for this experiment, performed on an Applied Chromatography Systems Model 750/03 constant-flow reciprocating pump, with Model 750/10A fixed-wavelength UV detector, are illustrated in figure 13.3 and summarised in table 13.4. Column and other conditions were as described above. The data in table 13.4 are for peak 2 (sodium phenobarbitone) only (V_R = 10,400 mm^3).

Using the equation above, an approximate estimate of the band dispersion (w_{lvi}) introduced by the loop valve injector is

$$w_{lvi}^2 = 1430^2 - 1200^2$$

whence w_{lvi} = 780 mm^3.

Comments

1. Clearly this experiment may be carried out on any HPLC system with suitable analytes. Indeed, this may well prove to be a profitable exercise in helping to establish the most suitable injection technique for a particular separation.

2. Further experiments may usefully examine the effect of injecting into porous PTFE frits (as compared with glass beads) and the effect of varying needle length or penetration (using spacers) on column performance.

3. The results, as those of experiment 13.1.1, were obtained with a column of very low efficiency. For a column giving, say, 5000 plates the effect of a valve dispersion amounting as here to 780 mm^3 would be catastrophic.

4. Experience with valves in general shows that with a really well-packed column there is little difference between plate efficiency with valve or syringe injection. The large additional dispersion found here is probably not due to the valve itself but to the fact that the sample is injected across the entire column section rather than into the core of the packed bed. (For further discussion see chapter 8.)

Acknowledgement

This experiment is based on a method developed by Whatman Lab-Sales Ltd.

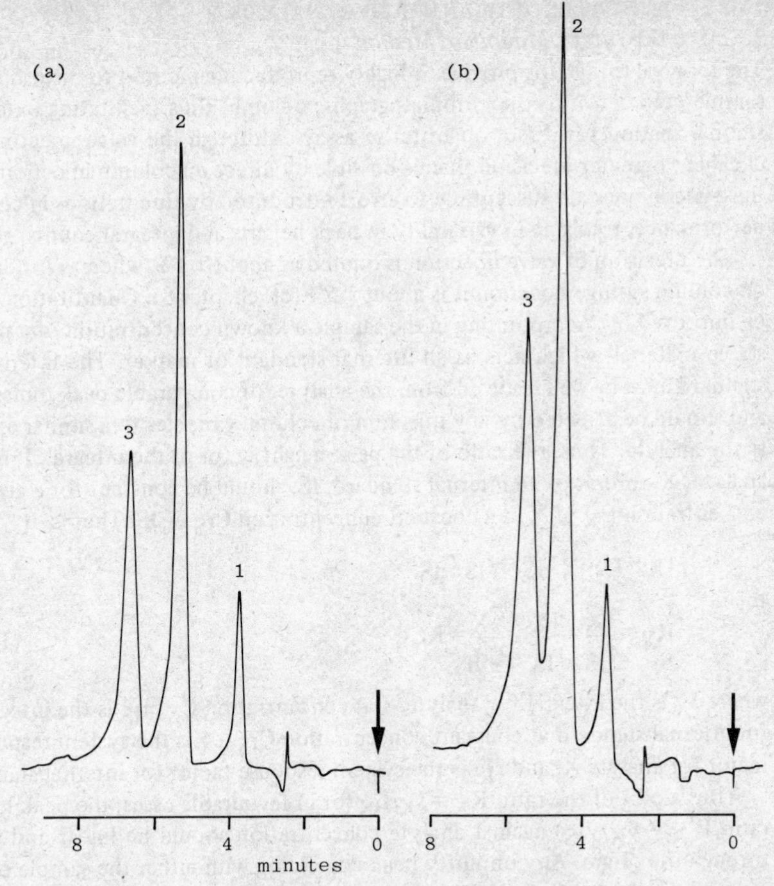

Figure 13.3. Effect of method of injection on resolution of barbiturates on Partisil ODS. UV detection, 254 nm, 0.04 AUFS (full details in text). (a) Syringe-septum injection, 3 mm^3; (b) Loop-valve injection, 3 mm^3.

Table 13.4. Effect of injection method.

	Experiment			
	A(3)	A(4)	A(5)	B
N	1200	1080	1200	850
w_{total} (mm^3)	1200	1300	1200	1430

13.2 PRINCIPLES OF QUANTITATIVE ANALYSIS

13.2.1 *The Internal Standard Method*

The loop valve injector provides a highly reproducible method for introducing sample volumes onto the chromatographic column, thus facilitating external standardisation for direct quantitative assay. Although the valve injector is capable of greater precision than is possible by direct on-column injection, both these techniques are susceptible to errors introduced by fluctuations in column performance, resulting in variability in peak heights and integral counts.

The precision of valve injection is quoted as about ± 1%, whereas for direct on-column syringe injection it is about ± 5% (cf. chapter 8). Quantitation may be improved by incorporating in the sample a known concentration of a pure stable material, which acts as an internal standard or marker. The internal standard must be well resolved from the analyte, of comparable peak dimensions, and should be affected by any injection or column variables to a similar extent as the analyte. Thus, the ratio of the peak height, y, or of the integral, I, for analyte, X, to that of an internal standard, IS, should be constant for a given concentration C_X of X in a constant concentration C_{IS} of IS. That is, if

$$I_X = r_X C_X \quad I_{IS} = r_{IS}\, C_{IS}$$

then

$$R_X = \frac{I_X}{I_{IS}} = \frac{r_X}{r_{IS}} \cdot \frac{C_X}{C_{IS}} = KC_X \qquad (1)$$

where I_X is the integral for analyte X at concentration C_X; I_{IS} is the integral for internal standard at constant concentration C_{IS}; r_X is the system response factor for analyte X; and r_{IS} is the system response factor for internal standard.

Thus a plot of the ratio $R_X = I_X/I_{IS}$ (or in favourable cases, the peak height ratio $R'_X = y_X/y_{IS}$) against analyte concentration should be linear and pass through the origin. Any impurity peak coinciding with either the sample or the internal standard peak should be estimated and the function I_X/I_{IX} compensated accordingly.

This method is capable of a precision better than ± 1% for on-column syringe injection, and is substantially independent of sample volume injected. In favourable cases a coefficient of variation of ± 0.5% or better has been reported (1).

If a UV detector is used, the main factors known to affect the precision of the internal standard method are:

(a) The selection of detector wavelength. Ideally both analyte and internal standard should exhibit broad maxima in the UV with closely similar λ_{max} values. In practice, a compromise wavelength must be found where both components absorb appreciably and where the slope $\Delta A/\Delta\lambda$ is low for each component. For analytes in low concentration the choice of wavelength may be restricted to the λ_{max} for the analyte to ensure maximum sensitivity.

(b) The absorbance range selected for the detectors. According to the well-known relationship between relative error and absorbance (shaped rather like a "bath tub"), the optimum absorbance, A, for measurement with minimum error is about 0.4 for a single-beam instrument, and about 0.8–1.2 for a double-beam spectrophotometer. Therefore where possible analyte concentrations should be adjusted to give an absorbance of 0.2–0.4 for optimum precision with single-beam instruments. However, the sample load required to achieve an absorbance of 0.4 at the analytical wavelength may lead to unacceptable deterioration in column performance. This is particularly so for low efficiency chromatographic systems, where peaks are heavily diluted, broad and difficult to detect. In these cases, a sensitive detector range (0.02 to 0.1 AUFS) might be required. It is precisely in this range where the relative error in absorbance measurement is highest, leading to lower precision in quantitative analysis.

(c) The concentration level of added internal standard. Ideally this should be adjusted so that at the observation wavelength the size of the internal standard peak falls in the centre of the calibration range. Ratios of I_X/I_{IS} greater than 10:1 and less than 1:10 are readily avoided by manipulation of sample and internal standard concentrations.

The present experiment is designed to show the principles of the internal standard method for a relatively simple analytical system (phenol in glycerin). A method for accurately incorporating the internal standard in the sample is described and the selection of a suitable analytical wavelength is illustrated. The use of peak height and area data is compared.

Equipment
Pump
Valve or septum injector
Variable-wavelength UV detector with 10 mm^3 flow-cell
Recorder
Integrator
Analytical glassware
Column System
100 × 5 mm packed with 5 μm SAS-Hypersil (a short alkyl side-chain bonded onto silica)
Mobile Phase
Methanol/Water (40:60 v/v)
Sample
phenol master solution in water 1000 μg cm^{-3}
p-cresol master solution in water 4000 μg cm^{-3} as internal standard
ampoules of Phenol in Glycerin BPC 5% w/v
Typical Operating Conditions
Detector wavelength: 271 nm
Detector sensitivity: 0.2 AUFS
Pressure drop across column: 60 bar (900 psi)
Mobile phase flow rate: 1.0 cm^3 min^{-1}

Experimental Procedure

1. Establish stable chromatographic conditions with the operating conditions described above.

2. Prepare the following standard mixtures of phenols by volumetric dilution of the two master solutions; in each case add to the volumetric flask exactly

one-tenth its volume of p-cresol master solution before making up to volume with distilled water:

500 μg phenol with 400 μg p-cresol per cm^3
400 μg phenol with 400 μg p-cresol per cm^3
250 μg phenol with 400 μg p-cresol per cm^3
100 μg phenol with 400 μg p-cresol per cm^3

3. Dilute the test injection with distilled water to give a final phenol concentration of about 250 μg cm^{-3} incorporating the p-cresol internal standard at a concentration of 400 μg cm^{-3} as before.

4. Adjust injection volume (ca. 4 mm^3) to give 60–80% full scale recorder deflection for the top standard (500 μg cm^{-3} phenol with IS). Adjust the integrator as required.

5. Inject each phenol standard in duplicate, recording the chromatogram and integrals.

6. Inject the test sample containing internal standard at least three times.

7. If time permits, select the phenol standard whose peak height ratio is closest to that of the test sample. Assay the test sample by the so-called "bracketing method" as follows:

(a) inject phenol standard
(b) inject test sample in duplicate
(c) inject phenol standard

8. If time permits, adjust the detector wavelength to 254 nm. Inject the 500 μg cm^{-3} phenol standard and record the chromatogram and integrals as before.

Calculation of Results

1. Tabulate the peak height and integral data. Calculate the peak height ratios and integral ratios for the phenol standards and plot each ratio against phenol concentration to produce two calibration graphs.

Evaluate the average peak height ratio and integral ratio for the test sample and interpolate its concentration on the appropriate calibration graph.

Comment on the linearity and y-intercept value for each calibration graph. Calculate the coefficient of variation (relative standard deviation) for the peak height ratios and the integral ratios recorded for the test sample.

2. Calculate the sample concentration from the "bracketing method" data as follows:

Average the peak height (or integral) ratio (R_S) for phenol standard C_S before and after test injection.

Average the peak height (or integral) ratio (R_X) for test sample dilution C_X.

Concentration of test sample dilution $C_X = \frac{R_X}{R_S} \times C_S$

Compare this concentration with that obtained by direct interpolation and comment.

N.B. a common variant of the bracketing method is to employ two standards, one slightly lower, the other higher than the test value. Duplicate injection of the lower standard is followed by the injections of test and finally duplicate

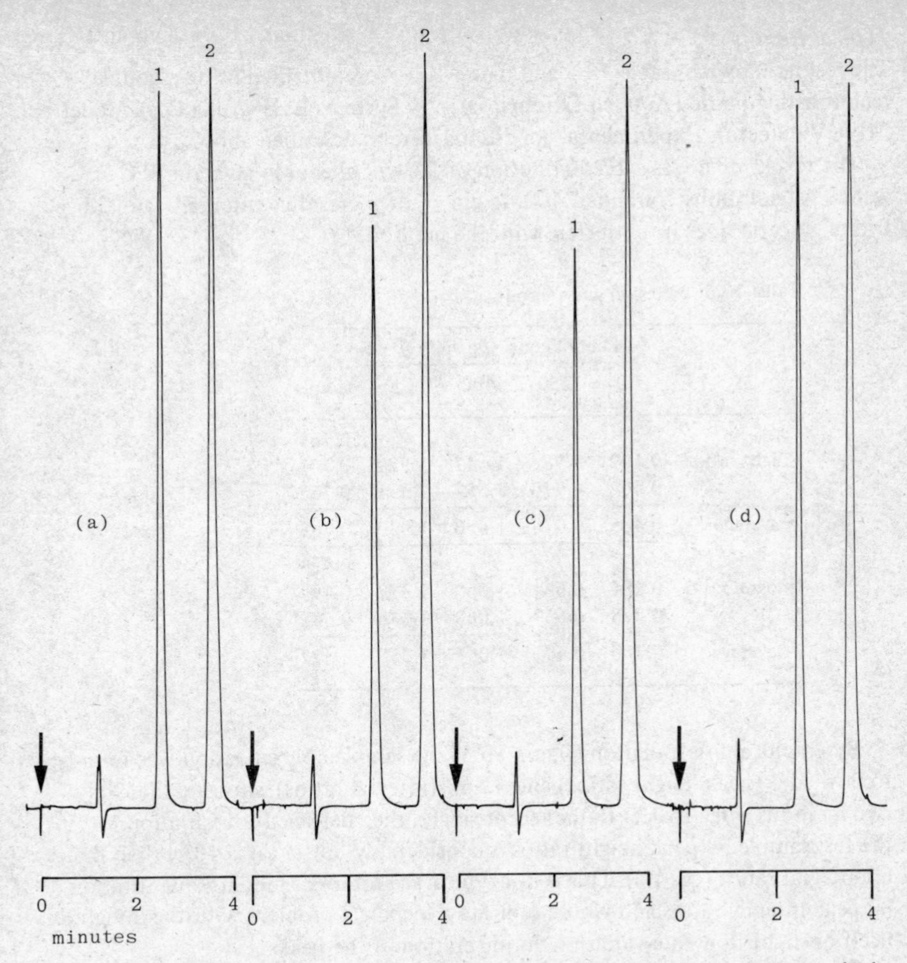

Figure 13.4. Quantitative analysis of phenol in glycerin by the bracketing method. Separation of (1) phenol and (2) p-cresol (internal standard) on SAS Hypersil. UV detection, 271 nm, 0.2 AUFS (full details in text). (a) and (d) 4 mm^3 injections of standard mixture containing 250 $\mu g\ cm^{-3}$ phenol and 400 $\mu g\ cm^{-3}$ p-cresol; (b) and (c) 4 mm^3 replicate injections of 250-fold diluted phenol-glycerin mixture containing 400 $\mu g\ cm^{-3}$ p-cresol.

injection of the higher standard. The test concentration is found by interpolation. This method is particularly useful for cases where the calibration graph is either non-linear or does not pass through the origin.

3. The test at 254 nm illustrates that the sensitivity of a quantitative method using a UV detector is wavelength dependent. The λ_{max} ϵ and $A^{1\%}_{1cm}$ values for pure phenol and p-cresol are 271 nm, 1520, 162, and 272 nm, 1363, 145 respectively, indicating that the optimum detector wavelength is 271 nm. This test may also be used to assess the sensitivity of the peak height (or integral) ratio to wavelength changes.

Typical Results

The results shown in table 13.5 and figure 13.4 were obtained using a double reciprocating pump (Applied Chromatography Systems Ltd) with a Cecil Model 212 UV detector. Experimental conditions were as described above.

The test sample was a 1:250 dilution of 5% w/v phenol in glycerin BPC in water. All solutions contained 400 μg cm^{-3} of p-cresol as internal standard. Dilute glycerin does not interfere with the method.

Table 13.5. Precision and reproducibility.

	Phenol conc. (μg cm^{-3})				test
	100	250	400	500	sample
					0.764
Height ratio	0.409	0.987	1.547	1.929	0.771
	0.402	0.970	1.552	1.931	0.768
average	0.405	0.978	1.550	1.930	0.768
					0.523
Integral ratio	0.284	0.684	1.060	1.327	0.499
	0.276	0.657	1.068	1.348	0.545
average	0.280	0.671	1.064	1.338	0.522

By graphical interpolation (figure 13.5) the sample concentration was found to be 4.91% (peak height ratios) and 4.86% (integral ratios), showing that the two methods agree to 1%. Rather surprisingly, the coefficient of variation for the test sample by peak height ratios is considerably better (at 0.46%) than that by integral ratios (ca. 4%). This is not typical for internal standard quantitation by peak integral ratios and would indicate a technical problem with the integrator itself or slight flow rate variation during elution of the peaks.

Typical data for an assay by the "bracketing method" yield the following concentrations for the same phenol in glycerin injection:

4.91% by peak height ratios and 4.79% by peak integral ratios.

Comments

1. Where high-performance chromatographic separation yields sharp, symmetrical peaks, the peak height ratio method is capable of a precision approaching ± 1% relative standard deviation.

2. The peak integral ratio method would be the method of choice for chromatograms where peaks were broad and tailing.

Acknowledgement

This method is based on a separation developed at the Wolfson Liquid Chromatography Unit, Edinburgh.

Reference

1. Fell, A.F., Kindlan, T. and Neil, J.M. (in preparation).

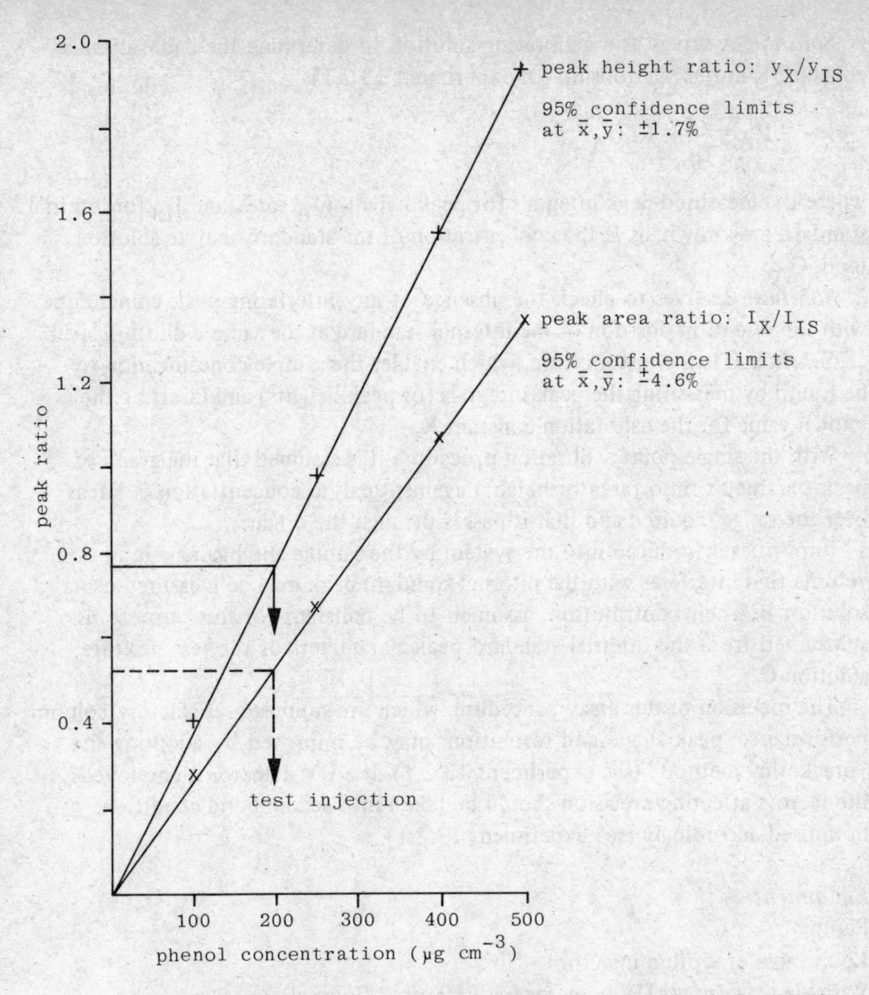

Figure 13.5. Calibration graph of peak height ratio and integral ratio for quantitation of phenol (X) in glycerin using p-cresol (IS) as internal standard. Conditions as for figure 13.4.

13.2.2 *Quantitation by the Single-Point Internal Standard Method*

This experiment describes a method of quantitation analogous to that used for British Pharmacopoeial assays by gas-liquid chromatography (1). In this procedure three solutions are prepared as follows:

A. known aliquot of standard analyte, X, plus known aliquot of internal standard solution, IS;

B. test sample extracted or dissolved in solvent as necessary, diluted to a suitable concentration for chromatography;

C. test sample diluted as for solution B, but incorporating the known aliquot of internal standard solution.

Equal volumes of each solution are injected; their function is described below:

Solution A serves as a calibrating solution to determine the calibration constant K in the relationship (of experiment 13.2.1)

$$R_A = \frac{I_A}{I_{IS}} = K.\,C_A \qquad (1)$$

where the measured peak integrals (or peak heights) I_A for X and I_{IS} for internal standard are known, as is the concentration of the standard analyte solution used, C_A.

Solution B serves to check the absence of any interfering peak coinciding with the retention position of the internal standard at the sample dilution level.

Solution C is the test mixture, which enables the sample concentration to be found by measuring the peak integrals (or peak heights) and inserting the known value for the calibration constant K.

With the single-point calibration procedure it is assumed that the graph of peak parameter ratio (area or height) against analyte concentration is linear over the range required and that it passes through the origin.

Impurities introduced into the system by the sample and having a retention volume that interferes with the internal standard peak may be measured using solution B. Their contribution, assumed to be constant for this purpose, is subtracted from the internal standard peak parameter for the test mixture, solution C.

The precision of this assay procedure, which pre-supposes satisfactory column performance, peak shape and resolution, may be improved by adopting the "bracketing method" (see experiment 13.2.1). If a UV detector is employed, the factors affecting precision should be taken into account and conditions optimised accordingly (see experiment 13.2.1).

Equipment
Pump
Loop valve or septum injector
Variable-wavelength UV detector with 10 mm^3 flow-cell
Recorder
Integrator
Analytical glassware

Column System
250 × 4.6 mm packed with 10 μm Partisil-ODS

Mobile Phase
Water/methanol/perchloric acid (55:45:0.2 v/v)

Sample
Methanol extract of rodenticide formulation containing Warfarin sodium: 2.00 g rodenticide extracted by 5.00 cm^3 methanol.
Warfarin sodium master solution: 400 $\mu g\ cm^{-3}$ in MeOH.
Tanderil (oxyphenbutazone) master solution as internal standard: 300 $\mu g\ cm^{-3}$ in MeOH.

Typical Operating Conditions
Detector wavelength: 300 nm

Detector sensitivity: 0.08 AUFS
Pressure drop: 95 bar (1400 psi)
Mobile phase flow rate: 2.0 cm^3 min^{-1}

Experimental Procedure

1. Establish stable chromatographic performance with the operating conditions described above.

2. Solution A: pipette 5.00 cm^3 of standard warfarin solution and mix with 5.00 cm^3 Tanderil solution. Inject 10 mm^3 and adjust detector sensitivity so that peaks are 60–80% full scale deflection (if necessary adjust volume appropriately). Set up the integrator as required.

3. Inject 10 mm^3 solution A and record the chromatogram and the integral for each peak.

4. Solution B: pipette 5.00 cm^3 of the sample warfarin extract and mix with 5.00 cm^3 methanol. Inject 10 mm^3 and record the chromatograms and peak integrals. Check whether any contaminant is detectable at the retention volume for the internal standard ($k' = 1.45$). If the warfarin peak is too large for the operating conditions selected, the volume mixed with 5.00 cm^3 internal standard may be reduced provided that the total volume for the mixture is maintained by addition of methanol to 10.00 cm^3 in a volumetric flask.

5. Solution C: pipette 5.00 cm^3 of the sample warfarin extract and mix with 5.00 cm^3 Tanderil solution. If necessary, pipette the smaller volume of test extract found in 4 above, mix with 5.00 cm^3 of Tanderil and make up to 10.00 cm^3 volumetrically. Inject 10 mm^3 and record the chromatogram and the peak integrals.

6. If time permits, repeat injections of solutions A and C in the sequence A-C-C-A, as in the bracketing method described in experiment 13.2.1.

Calculation of Results

1. Calculate the calibration constant K for the system, using data for solution A.

2. Calculate the concentration of warfarin sodium in test solution C and, assuming complete extraction, the amount in the rodenticide as percentage weight in weight. Correct the Tanderil integral of test solution C for any contaminant peak observed in the chromatogram for solution B.

3. Repeat the calculation, using the data obtained by the bracketing method:

$$C_X = \frac{R_X}{R_A} . C_A . F \qquad (2)$$

where R_A and R_X are the average integral (or height) ratios for solutions A and C respectively. C_A and C_X are the warfarin concentrations of standard solution and test sample extract respectively, and F is the dilution factor (relating the concentration of test solution C with the proportion of warfarin in the rodenticide on a percentage w/w basis, assuming complete extraction).

Typical Results

The results shown in table 13.6 and figure 13.6 were obtained using the Pye LC3 Chromatograph with variable wavelength detector and operating conditions as described above.

Table 13.6. Quantitation of Warfarin.

Solution	Warfarin peak height (mm)	Tanderil peak height (mm)	Peak height ratio R
A. (Calibrating)	181.5	220	0.825
B. (Check)	144	0	–
C. (Test)	140	212	0.660

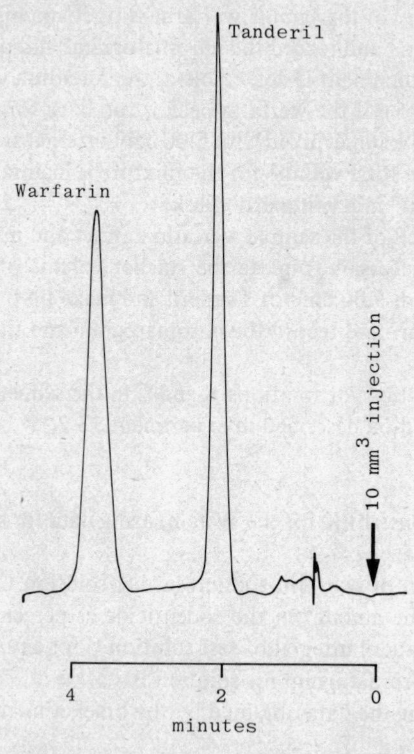

Figure 13.6. Quantitation of Warfarin using Tanderil as internal standard chromatographed on Partisil ODS. UV detection, 300 nm, 0.08 AUFS (full details in text).

Since the standard 400 μg cm^{-3} solution of warfarin is diluted 1:1 with Tanderil, the warfarin concentration C_X in the test sample solution is found using equation 1:

$$K = 0.825/200 = 0.004125 \text{ cm}^3 \mu\text{g}^{-1}$$

$$C_X = 0.660/0.004125 = 160 \mu\text{g cm}^{-3}$$

Alternatively, by equation 2:

$$C_X = \frac{0.660}{0.825} . 200 = 160 \ \mu g \ cm^{-3}$$

Thus the concentration of warfarin in the test extract (before dilution with internal standard 1:1) is 320 $\mu g \ cm^{-3}$. In the 5.00 cm^3 extract this represents 1.6 mg per 2 g rodenticide. The proportion of warfarin sodium is 0.8 mg per gram rodenticide, or 0.08% w/w, assuming complete extraction.

Comments

1. This method, although not official, could form the basis of a suitable HPLC procedure for the routine quality control of drugs, food additives and other materials such as the rodenticide in this example.

2. As illustrated in experiment 13.2.1, in developing a quantitative method it is advisable to establish that the calibration graph is linear over the required range and passes through the origin, before relying on a single concentration point calibration method.

3. The absorbance range suggested for this experiment (0.08 AUFS) could usefully be increased to ca. 0.2 AUFS with a consequent increase in quantitative precision (cf. experiment 13.2.1). This would require corresponding adjustment of the concentrations of solutions A, B and C.

4. The wavelengths of maximum absorbance for warfarin sodium and Tanderil in methanol are ~ 300 nm and 242 nm respectively. Comment on the selection of 300 nm as the analytical wavelength for this experiment.

Acknowledgement
This experiment is based on a method developed by Pye Unicam Ltd.

Reference

1. *British Pharmacopoeia* (1973) London.

13.3 DETECTION TECHNIQUES

13.3.1 *Sensitivity Enhancement and Wavelength Resettability in the Far UV*

Although the UV detector is generally useful for aromatic and conjugated organic molecules, many compounds have little if any useful absorptivity in the UV range from 220–340 nm. However, their absorptivity in the far UV (from 180–210 nm) may be considerable and usable to advantage for both qualitative and quantitative chromatographic analysis. With conventional UV detector technology, absorbance measurements in this region of the spectrum are notoriously unreliable. This is largely due to a combination of low spectral output for conventional deuterium sources, and in consequence the greater influence of stray light on measured absorbance. The low-volume flow cell (typically 10 mm^3) coupled with the fall-off in photomultiplier response below 220 nm puts even greater strain on the overall system.

Nevertheless, recent improvements in sources, optics and detector systems have led to the development of instruments that can operate satisfactorily in this region of the spectrum. This development has introduced a powerful detection capability for many compounds in addition to those that are otherwise transparent in the UV. This is because most aromatic and conjugated molecules have higher absorptivities in the far UV and may therefore be detected at correspondingly lower levels and quantified with greater precision. However, a problem arises in selecting solvents that are sufficiently transparent in the far UV for useful measurements to be made.

Another development in detector technology is the improvement in wavelength re-settability of monochromators. For internal standard methods where analyte and internal marker have differing λ_{max} values, it is helpful to be able to set the detector accurately to the analytical wavelength so that the calibration factor may remain constant (although this should be checked before and after an assay).

The present experiment is designed to illustrate the sensitivity enhancement in the far UV for a group of anticonvulsant drugs, two of which are almost transparent at the conventional detector wavelength of 254 nm (1).

The effect on peak height ratios when the analytical wavelength is varied by ± 0.5 nm is also examined.

Equipment

Pump

Loop valve injector

10 mm^3 flow-cell

Variable-wavelength detector with far-UV capability. Suitable spectrophotometers are the Perkin-Elmer Models LC-55 and LC-255, the Pye Unicam LC3 detector, the Varian Varioscan.

Recorder

Column System

250 × 2.6 mm packed with 9 μm HC-ODS-SIL X

Mobile Phase

Acetonitrile/water (25:75 v/v). (Note that doubly distilled, not deionised, water must be used for $\lambda < 220$ nm.)

Sample

Mixture of the following drugs, each at about 100 $\mu g\ cm^{-3}$ in eluent (listed in order of elution):

1. primidone
2. sodium phenobarbitone
3. carbamazepine

Typical Operating Conditions

Detector wavelengths: 195 nm, 208 nm and 254 nm

Detector sensitivity: 0.2 AUFS, 0.1 AUFS

Pressure drop: 200 bar (3000 psi)

Eluent flow rate: 1.0 $cm^3\ min^{-1}$

Experimental procedure

1. Establish stable chromatographic conditions with the operating data des-

cribed above.

2. Set the detector wavelength to 254.0 nm and inject 10 mm^3 of the sample mixture, recording the chromatogram at 0.2 AUFS. If necessary increase the detector sensitivity so that three distinct peaks are detected (k' values approximately 1.2, 2.4 and 7.0 for primidone, phenobarbitone and carbamazepine respectively).

3. Set the detector wavelength to 195.0 nm, inject 10 mm^3 sample mixture and record the chromatogram at 0.2 AUFS.

4. Set the detector wavelength to 208.0 nm and record the chromatogram for 10 mm^3 sample at 0.1 AUFS. Re-set the detector wavelength and repeat.

5. Repeat injections at 208.4 nm and 207.4 nm.

Calculation of Results

1. Calculate the sensitivity enhancement for each of the components detected at 195.0 nm relative to the response at 254 nm.

2. Calculate the peak height ratios of primidone and of phenobarbitone with respect to the carbamazepine peak at 207.4, 208.0 and 208.4 nm.

3. Assess the effect of detector wavelength re-settability on quantitative analysis for this mixture.

Table 13.7. Re-settability of wavelength.

	λ (nm)		
	207.4	208.0	208.4
Peak ratio 1 : 3	0.61	0.58, 0.58	0.56
Peak ratio 2 : 3	0.56	0.53, 0.54	0.52

Typical Results

1. At the 100 μg cm^{-3} level, primidone and phenobarbitone are barely detectable at 254 nm. Carbamazepine is detected with about 6 times greater sensitivity at 195 nm, as shown in figure 13.7.

2. Peak height ratios to carbamazepine (peak 3) for these particular concentrations vary with wavelength, as given in table 13.7.

Acknowledgement
This experiment is based on a method developed and published by Perkin-Elmer Ltd.

Reference

1. Adams, R.F., Vandemark, F.L., *Clin. Chem.* *22* (1976) 25–31.

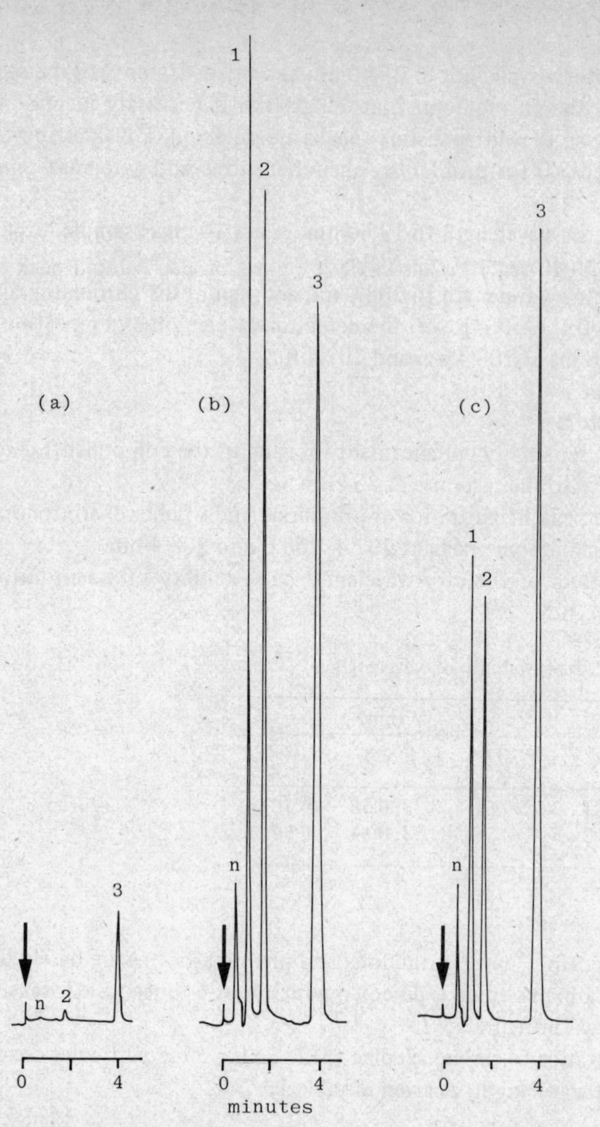

Figure 13.7. Effect of wavelength on detectability of anticonvulsants chromatographed on HC-ODS-Sil-X (full details in text). n denotes unretained solute. (a) 254.0 nm, 0.2 AUFS; (b) 195.0 nm, 0.2 AUFS; (c) 208.0 nm, 0.1 AUFS.

13.3.2 *Interrupted Elution Scan Techniques*

Confirmation of peak identity in HPLC often relies on the evidence of k′ values and solute behaviour in mobile phases of different polarity. Collection of eluted peaks affords an opportunity for microchemical identification tests to be applied but tends to be time-consuming and relatively insensitive unless high sample loading on column is employed. The availability of high-performance double-beam spectrophotometers for use with micro-litre flow cells offers the facility of scanning peaks routinely to obtain UV spectra for confirmation of identity and purity. Eluent flow is stopped during the scan, which should be rapid enough to avoid band broadening.

Stop-flow scanning enables optimum wavelengths to be established for detecting each component, thus increasing sensitivity and lowering detection limits.

The present experiment illustrates the use of this technique for a mixture of aromatic compounds. Repeated scans at three points on each eluted peak envelope enable an assessment of band purity to be made.

Equipment
Pump
Variable-wavelength double-beam spectrophotometer with 10 mm^3 flow-cell
Stop-flow valve for interrupted elution scanning
Septum injector
Recorder (or preferably two recorders) which can be used alternately for recording the chromatogram and the UV spectra

Column System
250 × 5 mm packed with 10 μm Hypersil

Mobile Phase
Hexane

Sample
A mixture of aromatic compounds in hexane (listed in order of elution):

1. bromonaphthalene 100 $\mu g\ cm^{-3}$
2. phenanthrene 100 $\mu g\ cm^{-3}$
3. azobenzene 400 $\mu g\ cm^{-3}$

Typical Operating Conditions
Detector wavelength: 254 nm for detection, variable 210–340 nm for identification
Detector sensitivity: 0.5 AUFS
Pressure drop: 70 bar (1000 psi)
Eluent flow rate: 5.0 $cm^3\ min^{-1}$

Experimental Procedure

1. Establish satisfactory base-line stability using the conditions suggested above.

2. Set the detector wavelength to 254 nm. Inject 5 mm^3 sample and record the chromatogram at a detector sensitivity of 0.5 AUFS.

3. Repeat the injection at reduced detector sensitivity (1.0 AUFS). As each peak maximum emerges, stop the flow and scan from 210–340 nm.

4. If time permits, step 3 may be extended by scanning each component at three points on the peak profile, e.g. 10% of peak maximum, at the peak maximum itself and just before the peak is completely eluted. (This enables peak purity to be assessed as discussed below.)

5. Having established the operational λ_{max} values for each component, set the detector to each wavelength in turn, inject 5 mm^3 sample and record the chromatograms at a detector sensitivity of 1.0 AUFS.

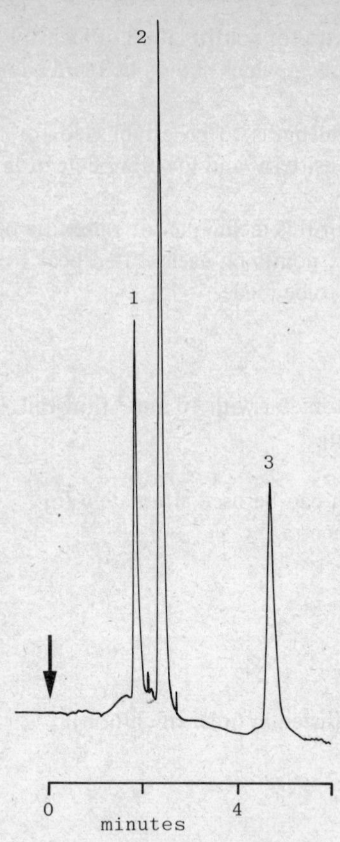

Figure 13.8. Chromatograms of (1) bromonaphthalene, (2) phenanthrene, (3) azobenzene on Hypersil. UV detection, 254 nm, 0.5 AUFS (full details in text).

Calculation of Results

1. Compare the interrupted elution spectra with reference spectra for each pure component and comment.

2. If repetitive scans have been carried out, comment on the apparent purity of each chromatographic peak.

3. Calculate the peak height ratios for components 1 and 2 relative to azobenzene in each of the chromatograms run at the three λ_{max} values for the components (bromonaphthalene λ_{max} = 220 nm; phenanthrene λ_{max} = 246 nm; azobenzene λ_{max} = 315 nm). Comment on the implications of this data for quantitative analysis and detection sensitivity.

Typical Results

The chromatogram illustrated in figure 13.8 and spectra illustrated in figure 13.9 were obtained using a Cecil Model 515 variable-wavelength detector with the Cecil Model 210 liquid chromatography system.

Acknowledgement

This experiment is based on a method developed by Cecil Instruments Ltd.

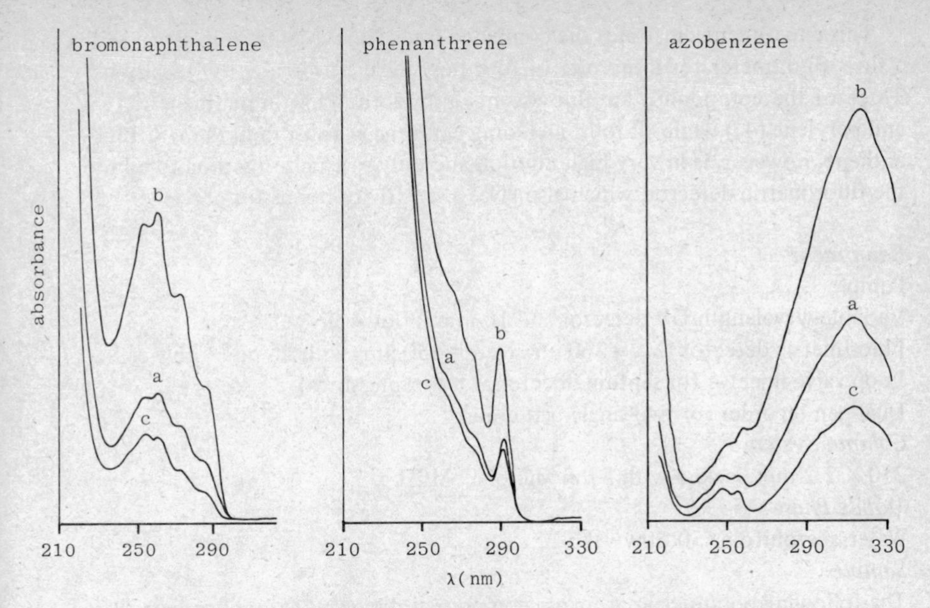

Figure 13.9. Interrupted elution scans of UV absorption spectra of the peaks of figure 13.8 at (a) 10% elution, (b) 50% elution, (c) 90% elution.

13.3.3 *Combined Fluorescence and UV Detection*

The use of a flow fluorimeter as an auxiliary or alternative detection method is well established, as discussed in chapters 10 and 11. A fairly large number of compounds may behave as fluorophores, which are generally detectable at 10–100 times lower concentrations than by their UV absorbance alone. Furthermore ability to fluoresce confers an extra degree of selectivity in itself. Other compounds may be converted, for example, to their dansyl or tosyl derivatives, or reacted with fluorescamine. Such procedures can be exploited with great effect for the sensitive detection of biological metabolites (1,2).

High sensitivity and precision in fluorimetric detection of eluted components requires judicious selection of primary excitation wavelength and secondary emission wavelength. The most common excitation source used is the medium-pressure mercury lamp, with a 366 nm primary filter. Other excitation wavelengths available are limited to the lines in the mercury emission spectrum (e.g. 436 nm) unless a high-intensity tungsten halide or xenon source is available. The sample fluorescence is monitored with an appropriate secondary filter to select the emission band maximum. These parameters may be optimised in situ, using the stop-flow technique described in experiment 13.3.2. Equipment is now becoming available to scan the fluorimetric emission spectrum of components separated by HPLC and, of course, the UV absorption or excitation spectrum may be readily scanned, as described in chapter 10. Different flow-cells are used for UV and fluorescence detectors; the latter generally incorporate some device to allow lateral observation of emitted fluorescence radiation (see figure 10.6) in order more easily to prevent flooding of the detector by the excitation radiation.

This experiment illustrates the combined use of a UV detector in series with a flow fluorimeter for a mixture of four polynuclear aromatic hydrocarbons. Three of the compounds are fluorescent (anthracene (1), fluoranthene (2) and perylene (4)) while all four, including chrysene (3) absorb in the UV. Fluoanthene, however, is in very high dilution and only satisfactorily monitored by the fluorimetric detector, which also gives a sensitivity bonus for perylene.

Equipment
Pump
Variable-wavelength UV detector and 10 mm^3 flow-cell
Fluorimetric detector (λ_{ex} = 360 nm, λ_{em} = 450 nm) with 25 mm^3 flow-cell
Loop valve injector (or septum injector at lower pressures)
Dual-pen recorder (or two single-pen units)
Column System
250 × 2.2 mm packed with 5 μm MicroPak-MCH
Mobile Phase
Water/acetonitrile (50:50 v/v)
Sample
The following polynuclear aromatic compounds dissolved in mobile phase (listed in order of elution):

(A) 1. anthracene 10 μg cm^{-3}
2. fluoranthene 0.5 μg cm^{-3}
3. chrysene 10 μg cm^{-3}
4. perylene 10 μg cm^{-3}
(B) mixture A diluted 1:10 with mobile phase.
(C) perylene (10 μg cm^{-3}) in mobile phase.

Typical Operating Conditions
UV-detector wavelength: 254 nm
UV-detector sensitivity: about 0.1 AUFS
Fluorimeter excitation wavelength: 360 nm
Fluorimeter emission wavelength: 450 nm
Pressure drop: 170 bar (2500 psi)
Mobile phase flow rate: 1.5 cm^3 min^{-1}

Special Features
Column eluent flows first through the UV detector flow-cell, which is connected in series with the fluorimeter flow-cell to enable simultaneous monitoring of both parameters.

Experimental Procedure

1. Establish stable chromatographic conditions using the operating parameters suggested above. A solution of 10 μg cm^{-3} anthracene may be injected (10 mm^3) to set the UV detector sensitivity to give about 80% full scale deflection. Similarly, a solution of 10 μg cm^{-3} perylene may be employed to set fluorimeter sensitivity. If a dual-pen recorder is used, it should be arranged that the pens record the UV and fluorimetric detector outputs in opposite directions.

2. Inject 10 mm^3 solution A and record the two detector outputs simultaneously. Identify the components in each chromatogram.

3. Increase each detector sensitivity by a factor of 10 and inject 10 mm^3 of solution B, recording the detector outputs simultaneously as before.

4. If time permits, dilute solution C step-wise with mobile phase by factors of 10, inject 10 mm^3 quantities and increase fluorimeter sensitivity progressively to determine the minimum detectable quantityof perylene.

Calculation of Results

1. Identify the four components in the chromatograms for the two detectors. The UV detector picks up components, 1, 3 and 4, while the fluorimetric detector records peaks 1, 2 and 4 only.

2. Calculate the sensitivity ratios for sample mixture A by each method of detection (relative to the smallest peak in each chromatogram).

3. Calculate the minimum detectable quantity for perylene, assuming that the detection limit is equivalent to twice the peak-to-peak noise (i.e. a signal:noise ratio of 2:1).

Typical Results

The data shown in table 13.8 and figure 13.10 were obtained using the Varian 8510 isocratic chromatograph equipped with a Varichrom variable-wavelength UV detector and a Fluorichrom fluorescence detector. Operating conditions were as described above.

Table 13.8. Relative peak heights.

	anthracene	fluoranthene	chrysene	perylene
UV detector	25	0	6	1
fluorimeter	3	1	0	15
concentration ratios	20	1	20	20

1. Typical capacity factor (k') values are: anthracene 1.50, fluoranthene 2.20, chrysene 3.80, perylene 5.30.

2. For mixture A, the relative peak heights are typically as shown in table 13.8.

Anthracene, although fluorescent, is seen to be more sensitively detected by its UV absorption. By contrast, fluoranthene is a very strong fluorophore and readily monitored by the flow-fluorimeter detector at a concentration well below that for UV detection.

Chrysene is not fluorescent, but is satisfactorily monitored by the UV detector at this concentration level.

Perylene is almost as strong a fluorophore as fluoranthene under these fluorimeter conditions. As compared with the performance of the UV detector, fluorimetry would be the method of choice for monitoring and quantitation.

3. The minimum detectable quantity of perylene would depend upon the range and versatility of the fluorimeter employed. In this experiment it was

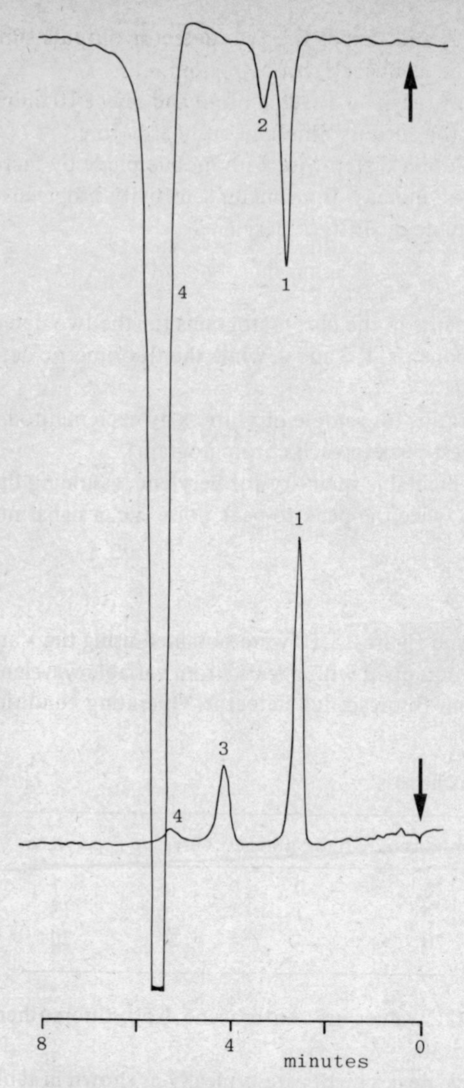

Figure 13.10. UV followed by fluorimetric detection in chromatography of aromatic hydrocarbons on MicroPak-MCH (full details in text): Lower trace: UV detection, 254 nm, 0.02 AUFS. Upper trace: fluorimetric detection, excitation 360 nm, emission (detection) 450 nm.

about 2 ng (10 mm^3 injected of a 0.2 $\mu g\ cm^{-3}$ solution). The minimum detectable quantity of fluoranthene was about 1 ng.

Comments

1. These results show both the selectivity and the sensitivity of fluorimetric detectors as compared with the conventional UV monitor. The relevance of fluorimetric detectors for low level detection, for example in environmental

analysis, is illustrated by this particular experiment. However, the potential range of applications is limited both by the occurrence of natural fluorescence in compounds and by the availability of suitable excitation and emission conditions in current LC flow-cell fluorimeters.

2. The geometry of the flow cells is crucial in determining the extent of band broadening (cf. experiment 13.1.3). If unsatisfactory band broadening is observed when using both flow cell detector systems in series, the experiment should be performed with each detector singly.

3. Fluorimetric detectors capable of scanning the emission spectrum under stop-flow conditions are now available commercially (e.g. the Perkin-Elmer model LC1000 and series 204). They enable further spectral information to be obtained for identification purposes.

4. The linearity of response to concentration in fluorimetry is notoriously sensitive to quenching phenomena. The excited fluorophore may lose its energy to quenching agents in the eluent (e.g. chloride anions), or to ground state fluorophores. This necessitates the most careful check on detector response over the concentration range of analytical interest if quantitation is to be successfully performed. In any event, the UV absorbance should not exceed 0.1 A for the relationship $F = kc$ to hold (see section 10.3(b), equations 10.3, 10.4), where F is the fluorescence (usually measured as a percentage of full scale deflection), c is the concentration (in convenient units), and k is an instrumental constant embracing incident light intensity (I_0), quantum yield of fluorescence, absorptivity of fluorophore.

5. Although this particular mixture is amenable to separation under isocratic conditions, more complex samples may require a gradient elution method (cf. experiments 13.4.1 and 2).

Acknowledgement
This experiment is based on a procedure devised by Varian Associates Ltd.

References
1. Udenfriend, S., *Fluorescence Assay in Biology and Medicine,* Vol I and II. Academic Press, 1969.
2. Udenfriend, S. *et al., Science 178* (1972) 871.

13.4 GRADIENT ELUTION TECHNIQUES

13.4.1 *Isocratic Compared with Gradient Elution: Insecticide Intermediates*

Isocratic conditions of operation, where the composition of the mobile phase is maintained constant, offer the most suitable conditions for quantitative anlaysis by LC, where stability and especially base-line stability are crucial (cf. experiment 13.2.1). In a sense, this mode of operation is comparable with isothermal conditions in gas-liquid chromatography. However, where a mixture of components with widely differing elution properties on a given packing material has to be separated, analysis time may be long and later peaks in the chromatogram are broadened. Just as temperature programming is used in GLC to resolve wide-range boiling-point mixtures and to shorten analysis time, so in HPLC the ratio of two (rarely more) mobile phase constituents may be varied to produce a programmed change in eluent polarity. This technique, known as gradient elution (cf. chapter 9), may employ various types of programme to produce a particular profile of mobile phase composition. Gradient elution, or solvent programming, not only shortens analysis time, but enables separations to be achieved in one chromatogram that would otherwise require two or even more sets of isocratic conditions to be set up. Gradient elution also offers the facility of rapid "scouting" of solvent mixtures to establish optimum isocratic conditions more easily.

The present experiment illustrates the separation of five insecticide intermediates by gradient elution. This technique is compared with the results obtained under various isocratic conditions.

Equipment

Pump system with gradient capability (or low-pressure solvent mixing system and pump with low swept volume)
Variable-wavelength UV detector with 10 mm^3 flow-cell
Septum or loop valve injector
Recorder

Column System

250 × 4.6 mm packed with 10 μm LiChrosorb SI 60 (packed by stirred 5% ammonium slurry).

Mobile Phase

Hexane/dichloromethane proportioned to give progressive increase in dichloromethane from 5–100% v/v for gradient elution.

Sample

A mixture of five insecticide intermediates, each at 100 μg cm^{-3} in methanol (listed in order of elution):

1. 3,4-dichloronitrobenzene
2. 2,3-dichloronitrobenzene
3. 2,5-dichloroaniline
4. 2,3-dichloroaniline
5. 3,4-dichloroaniline

Typical Operating Conditions

Detector wavelength: 285 nm
Detector sensitivity: 0.1 AUFS
Pressure drop: 25 bar (350 psi)
Mobile phase flow rate: 2.0 cm^3 min^{-1}

Special Features

Various types of solvent programme have been developed for LC in order to produce a variety of eluent composition profiles. In the present experiment, a "concave" profile was used to obtain the results illustrated.

Experimental Procedure

1. Establish stable isocratic chromatographic conditions for hexane/dichloromethane (90:10 v/v). Inject 6 mm^3 of a 100 μg cm^{-3} solution of 3,4-dichloro-

nitrobenzene to set detector sensitivity to 80% full scale deflection.

2. Inject 6 mm^3 of the sample mixture under these isocratic conditions. When four of the components have eluted (ca. 10 minutes) run a solvent gradient up to 50% dichloromethane in order to elute rapidly the strongly retained fifth component.

3. Inject 6 mm^3 sample under isocratic conditions with solvent composition hexane/dichloromethane (50:50 v/v) noting that only four of the five components are resolved.

4. Establish stable isocratic conditions for hexane/dichloromethane (95:5 v/v). Set the solvent programmer to generate a "concave" gradient of dichloromethane composition in the mobile phase from 5–100% in 8 minutes, i.e. a slow initial rate of increase in dichloromethane concentration, followed by a rapid increase towards the end of the 8-minute programme cycle.

5. Inject 6 mm^3 of sample, start the gradient and record the chromatogram.

6. If possible, superimpose on the resultant chromatogram the profile of the gradient generated.

Calculation of Results

1. Calculate the retention times for:
(a) components 1–4 under the 90:10 v/v isocratic conditions
(b) components 1, 3 and 4 and 5 under the 50:50 v/v isocratic conditions
(c) all five components under gradient elution conditions

2. If possible, in the gradient elution chromatogram, measure the percentage composition of the mobile phase at the retention times for each peak.

3. Compare the results from each chromatogram. It will be observed that for a low percentage of dichloromethane, reasonable resolution of components 1 and 2 is achieved, at the cost of a long retention time for peak 5 (ca. 23 minutes; cf. figure 13.11 a).

Increasing the eluent polarity shortens analysis time considerably but resolution of peaks 1 and 2 is sacrificed (figure 13.11 b).

Gradient elution under these conditions maintains resolution with considerable reduction in analysis time (figure 13.11 c).

Typical Results

Typical chromatograms are shown in figure 13.11; retention times are listed in table 13.9. The data were obtained on a Micromeritics Model 7000 B with concave gradient type N5 using the chromatographic conditions described above.

Comments

The eluent programme profile used to obtain these results is by no means the only solution to the problem of ensuring adequate resolution for early components, while reducing analysis time. For example, an isocratic period of about 5 minutes followed by a more or less linear solvent programme would be expected to yield satisfactory results for this mixture.

Acknowledgement

This experiment is based on a method developed by Coulter Electronics Ltd.

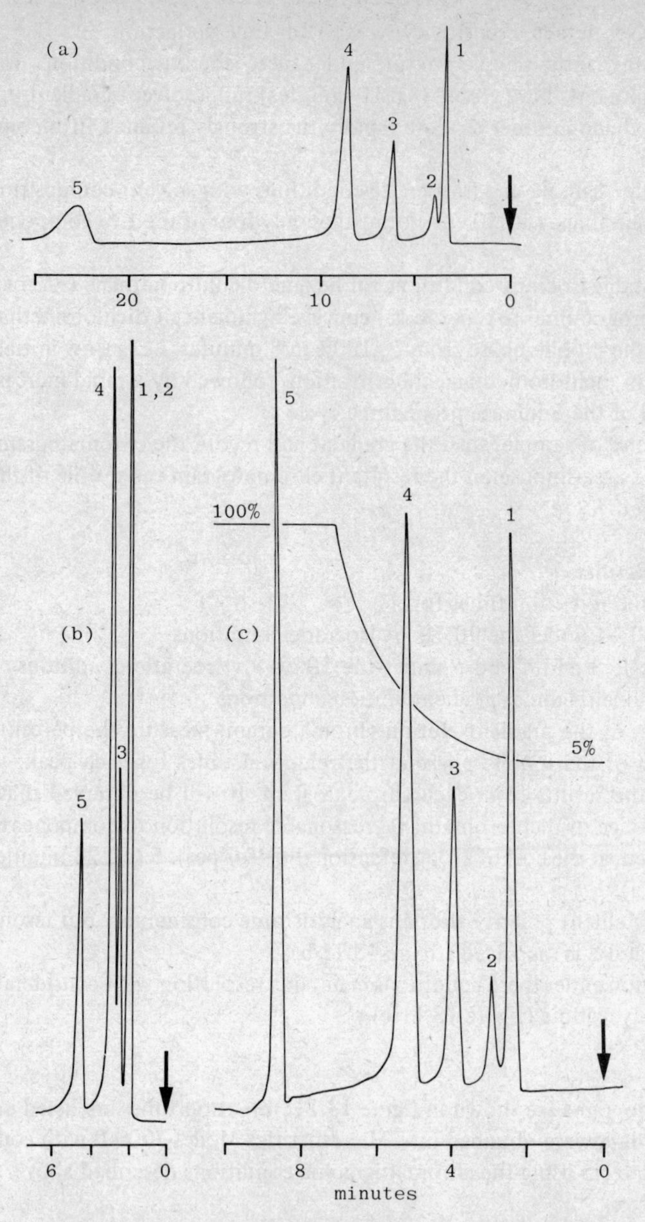

Figure 13.11. Isocratic and gradient elution of insecticide intermediates in LiChrosorb SI 60. UV detection, 285 nm, 0.1 AUFS (full details in text). (a) isocratic elution 90:10 v/v, hexane/methylene chloride; (b) isocratic elution 50:50 v/v, hexane/methylene chloride; (c) gradient elution with concave gradient from 95:5 to 0:100 v/v hexane/methylene chloride as eluent.

Table 13.9. Elution times in minutes for various hexane/dichloromethane eluent compositions.

	Eluent composition			
	99:10	50:50	Concave gradient	
	t_R	t_R	t_R	%DCM
1. 3,4-dichloronitrobenzene	3.4	2.0	4.5	9
2. 2,3-dichloronitrobenzene	3.9	2.0	5.4	15
3. 2,5-dichloroaniline	6.0	2.8	8.0	30
4. 2,3-dichloroaniline	8.5	3.1	9.0	70
5. 3,4-dichloroaniline	22.6	5.4	10.0	100

13.4.2 *Isocratic compared with Gradient Elution: sulphonamides*

An alternative experiment to 13.4.1 shows the effect of gradient elution on the separation of a mixture of sulphonamides. Isocratic chromatograms are run to compare both the resolution and the quantitative stability of the two systems.

Equipment
Pump system with gradient elution capability (or low-pressure solvent-mixing system and pump with low swept-volume)
Fixed or variable-wavelength UV detector with 10 mm^3 flow-cell
Loop valve injector (20 mm^3 volume)
Recorder
Integrator
Column System
250 × 4.5 mm packed with 5 μm Spherisorb S5-ODS
Mobile Phase
Water/acetonitrile proportioned to give progressive increase in acetonitrile from 8–25% v/v for gradient elution.
Sample
A mixture of sulphonamides in methanol (listed in order of elution):
1. Sulphadiazine 220 $\mu g\ cm^{-3}$
2. Sulphamerazine 100 $\mu g\ cm^{-3}$
3. Sulphamethoxazole 50 $\mu g\ cm^{-3}$
4. Sulphaquinoxaline 70 $\mu g\ cm^{-3}$

Typical Operating Conditions
Detector wavelength: 254 nm
Detector sensitivity: 0.1–0.2 AUFS
Pressure drop: 75 bar (1100 psi)
Mobile phase flow rate: 2.0 $cm^3\ min^{-1}$

Special Features
As in experiment 13.4.1, a concave gradient profile was used.

Expenmental Procedure

1. Establish stable isocratic chromatographic conditions for water/acetonitrile (75:25 v/v). Inject 20 mm^3 of a 100 $\mu g\ cm^{-3}$ solution of sulphamerazine to adjust detector sensitivity to give 80% full scale deflection.
2. Inject 20 mm^3 of the sample mixture under these isocratic conditions and record the chromatogram, integrating each peak area.
3. Repeat 2 at least twice.
4. Adjust the eluent composition to give water/acetonitrile (92:8 v/v) and establish base-line stability. Select a "concave" gradient elution programme for 8–25% acetonitrile in 8 minutes.
5. Inject 20 mm^3 sample, start the gradient and record the chromatogram.
6. Superimpose the solvent profile on the chromatogram.

Calculation of Results

1. Calculate the retention times for all four components under (a) isocratic conditions (b) gradient elution conditions.
2. Calculate the peak height (or peak integral) ratios for the repeated isocratic chromatograms, with reference to peak 4 (sulphaquinoxaline) as internal standard.
3. Measure the extent of base-line drift in the gradient elution chromatogram and assess its effect on quantitative analysis for this mixture if peak 4 were taken as the internal standard. Suggest what improvements could be devised for successful quantitation by gradient elution chromatography.

Typical Results

Typical chromatograms are shown in figure 13.12 and retention times in table 13.10. Quantitation is shown to be good in table 13.11.

Table 13.10. Retention times in minutes for isocratic vs gradient elution.

	Isocratic 75:25 v/v	Gradient elution 8-25% acetonitrile
Sulphadiazine	1.7	3.6
Sulphamerazine	1.9	4.8
Sulphamethoxazole	2.9	8.0
Sulphaquinoxaline	4.2	10.6

Table 13.11. Peak height ratios relative to sulphaquinoxaline.

	1	2	3	4	standard deviation	CV%
Sulphadiazine	0.352	0.354	0.353	0.347	0.003	0.9
Sulphamerazine	1.091	1.097	1.084	1.085	0.006	0.6
Sulphamethoxazole	0.644	0.644	0.644	0.644	–	–

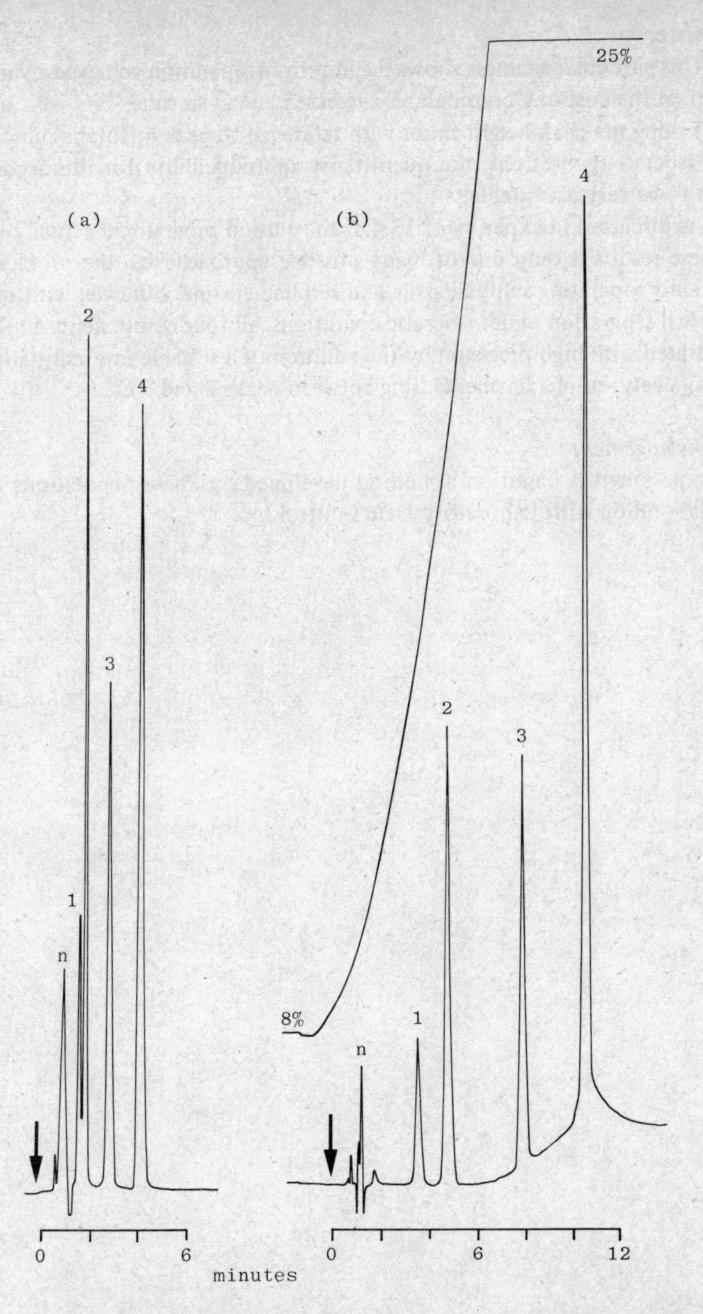

Figure 13.12. Isocratic and gradient elution of sulphonamides on Spherisorb ODS. UV detection, 254 nm, 0.128 AUFS. n denotes unretained solute (full details in text). (a) isocratic elution 25:75 v/v, acetonitrile/water; (b) gradient elution with concave gradient from 8:92 to 25:75 v/v acetonitrile/water as eluent.

Comments

1. This particular example shows the improved separation achieved by gradient elution, at the cost of a considerable increase in analysis time.

2. Taking the peak height ratios with reference to peak 4 (sulphaquinoxaline) under isocratic conditions, the quantitative reproducibility for this separation is seen to be very acceptable.

3. As discussed in experiment 13.4.1, the elution programme actually used for these results is only one of many possible approaches to the problem of adequately separating sulphadiazine and sulphamerazine. But even with the less than ideal separation under isocratic conditions, all four components could be quantitated with high precision by the addition of a suitable internal standard (e.g. N_4-acetyl-sulphadiazine, eluting between peaks 2 and 3).

Acknowledgement

This experiment is based on a method developed by Phase Separations Ltd in collaboration with Laboratory Data Control Inc.

APPENDIX. PROPERTIES OF CHROMATOGRAPHIC SOLVENTS

Appendix. Properties of chromatographic solvents.

	ε^o	η	δ	A_s	MW	ρ	c	RI	BP	λ	
	(a)	(b)	(c)	(d)	(e)	(f)	(g)	(h)	(i)	(j)	
perfluoroalkanes	-0.25	0.39	6.0		340	1.70	5.0	1.267	57.0	210	Fluorinert FC78 (3M), Flutec PP1 (ISC) mainly perfluror-n-hexane, expensive.
1,1,2-trichlorotrifluorethane		0.69	7.0		187.40	1.517	8.1	1.250	47.6	230	Arklone P (ICI) stabilizer free.
pentane	0.00	0.23	7.1	5.9		0.629	8.7	1.358	36.2	210	Low bp, more expensive but less toxic than hexane, miscible with CH_3CN in all proportions.
2,2,4-trimethylpentane	0.01	0.50		7.6	114.20	0.692	6.1	1.404	99.3	210	Less aromatics, higher bp and more viscous than hexane, more miscible.
hexane	0.01	0.33	7.3		86.17	0.659	8.0	1.375	86.2	200	Good UV.
cyclohexane	0.04	1.00	8.2	6.0	84.16	0.779	9.3	1.427	81.4	210	Rather viscous but different selectivity from hexane.
cyclopentane	0.05	0.47		5.2	70.14	0.740	10.7	1.406	49.3	210	Expensive.
1-pentene	0.08	0.24		5.8	70.14	0.640	9.3	1.371	30.0	215	Not readily available.
carbon disulphide	0.15	0.37	10.0	3.7	76.14	1.260	16.6	1.626	46.2	380	Too smelly and toxic, not UV.
carbon tetrachloride	0.18	0.97	8.6	5.0	153.80	1.590	10.4	1.466	76.8	265	Rather viscous, poor UV.
pentyl chloride	0.26	0.43	9.0	4.2	106.60	0.883	8.3	1.413	108.2	225	
m-xylene	0.26	0.62	8.9	7.6	106.17	0.864	8.2	1.500	139.1	290	Not UV.
propyl ether	0.28	0.37	7.3	5.1	102.20	0.747	7.1	1.368	89.64	220	Peroxide hazard, rather low boiling.
2-chloropropane	0.29	0.33	8.4	3.5	78.54	0.862	10.9	1.378	35.74	225	Too expensive, 1-chlorobutane similar.
toluene	0.29	0.59	8.9	6.8	92.13	0.867	9.4	1.496	110.6	285	Not UV.
1-chlorobutane	0.30	0.35	8.4		92.57	0.873	10.0	1.397	68.3	225	Expensive, t-butylchloride cheaper.
chlorobenzene	0.30	0.80	9.5	6.8	112.56	1.106	9.8	1.525	131.7		Not UV.
benzene	0.32	0.65	9.2	6.0	78.12	0.879	11.3	1.501	80.1	280	Not UV.
diisopropyl ether		0.329	7.0		102.19	0.724	7.1	1.368	68.3		Peroxide problems.
ethyl ether	0.38	0.23	7.4	4.5	74.12	0.713	9.6	1.353	34.6	220	Peroxide hazard, too low boiling.
ethyl sulphide	0.38	0.45	8.9	5.0	90.19	0.836	9.3	1.442	92.1	290	Not UV.
chloroform	0.40	0.57	9.3	5.0	119.39	1.500	12.6	1.443	61.2	245	UV not ideal.
methylene chloride	0.42	0.44	9.7	4.1	84.94	1.336	15.7	1.424	40.1	235	UV not ideal.
tetrahydrofuran	0.45	0.55	9.1	5.0	72.10	0.880	12.3	1.408	66.0	215	Peroxide problems.
1,2-dichloroethane	0.49	0.79	9.8	4.8	96.95	1.250	12.7	1.445	83.0	225	Best UV of chlorinated solvents.
methylethylketone,butan-2-one	0.51	0.40	9.3	4.6	72.11	0.805	11.2	1.381	79.6	330	Not UV.
1-nitropropane	0.53	0.84	9.9	4.5	89.10	1.001	11.2	1.400	131.2	380	Not UV.
acetone	0.56	0.32	9.9	4.2	58.08	0.818	12.6	1.359	56.5	330	Not UV but good solvent and cheap.
dioxan	0.56	1.54	10.0	6.0	88.11	1.033	11.7	1.422	101.3	220	Purification needed to achieve good UV.
ethyl acetate	0.58	0.45	9.6	5.7	88.10	0.901	10.2	1.370	77.15	260	UV not ideal.

methyl acetate	0.60	0.37	9.2	4.8	74.08	0.927	12.5	1.362	57.1	260	UV not ideal.
pentanol	0.61	4.10	9.8	8.0	88.15	0.815	9.2	1.410	137.8	210	
dimethylsulphoxide	0.62	2.24	12.0	4.3	78.13	1.100	14.0	1.478	189.0		Much too viscous except as a minor component.
aniline	0.62	4.40	10.3	6.7	93.13	1.022	11.0	1.586	184.4		
diethylamine	0.63	0.38	8.0	7.5	73.14	0.702	9.7	1.387	55.5	275	Limited UV, but basic.
triethylamine		0.38	7.5		101.10	0.728	7.2	1.401	89.5		Basic.
nitromethane	0.64	0.67	12.6	3.8	61.04	1.138	18.5	1.394	101.2	380	Expensive, not UV.
acetonitrile	0.65	0.37	11.7	10.0	41.05	0.782	19.1	1.344	82.0	190	Expensive but otherwise ideal.
pyridine	0.71	0.94	10.7	5.8	79.10	0.983	12.4	1.510	115.3	305	Very limited UV, but basic.
dimethoxyethane		0.45			90.12	0.863	9.5	1.380	83.0	220	
2-methoxy ethanol	0.74	1.72	11.4	6.3	76.10	0.965	7.7	1.401	124.6	220	Good UV if pure enough.
propan-2-ol	0.82	2.30	11.5	8.0	60.10	0.786	13.4	1.380	82.3	210	More viscous than ethanol and methanol, but popular additive.
ethanol	0.88	1.20	12.7	8.0	46.07	0.789	17.1	1.361	78.5	210	Excise duty problems; more viscous than methanol but more miscible.
methanol	0.95	0.60	14.4	8.0	32.04	0.796	24.9	1.329	64.7	205	Cheap and excellent UV if pure enough.
ethane-1,2-diol	1.11	19.90	14.6	8.0	62.07	1.114	18.0	1.427	197.3	210	Much too viscous.
acetic acid	Large	1.26	10.1	8.0	60.05	1.049	17.5	1.372	117.9		Useful additive-acid.
water	Large	1.00	23.4		18.00	1.000	55.6	1.330	100.0		
Solvents for Exclusion Chromatography											
o-dichlorobenzene (ODCB)		1.324	10.0		147.00	1.306	1.32	1.552	180.48		Used hot (typically 130°C).
1,1,2-trichloroethylene		0.566	9.2		131.40	1.460	1.11	1.476	87.19		
N,N-dimethylformamide		0.92	12.1		73.10	0.949	1.30	1.428	153.0		
tetrachloroethylene		0.90	9.3		165.80	1.620	0.98	1.505	121.0		
N-methylpyrrolidone		1.65	11.3		86.15	0.819		1.470	202.0		
1-methylnaphthalene					142.19	1.025	0.72	1.618	235.0		Used hot (typically 165°C)
dimethylacetamide					87.12	0.927	1.07	1.438	165.0		
trans-decahydronaphthalene (decalin)					138.25	-0.896	0.648	1.470	194.6		
trifluoroethanol		1.20			100.04	1.390	1.39	1.219	73.6		

Notes to Appendix

(a) ε^{o} = values for alumina according to Snyder (ref. 3 of Chapter 3).

(b) η = viscosity, in centipoise or mN s m^{-2} at 20°C.

(c) δ = solubility parameter, in (cal $cm^{-3})^{1/2}$.

(d) A_s = molecular area, in units of 0.085 nm^2 (see Chapter 3).

(e) MW = molecular weight, in g mol^{-1}.

(f) ρ = density, in g cm^{-3}.

(g) c = molarity of liquid = $1000\rho/MW$, in mol dm^{-3}.

(h) RI = refractive index at 20°C.

(i) BP = boiling point in °C at 1 atm pressure.

(j) λ = UV cut-off: wavelength in nm at which transmission falls to 10% in a 10 mm cell.

Acknowledgment: The data in this table are taken with minor changes from *Liquid Chromatography in Practice* by Paul A. Bristow, publ. HETP, Wimslow, Cheshire, 1976.

ADDENDA AND CORRIGENDA

p.3	line 25 should read "... but in practice it is usually..."
p.6	in equation 2.6, for "t_r" read "t_R"
p.13	in figure 2.5, the three chromatograms should be labelled, from the top, "(a)", "(b)", and "(c)"
p.29	in table 3.4, line 5, for "Reeve Angel" read "Whatman"
p.50	line 13 should read "... plate-height/velocity curves..."
p.63	caption to figure 6.7 should include reference citation "(24)"
p.134	in figure 11.2, for "Fluram" read "fluorescamine", and in caption, for "(LVP)" read "(LYP)"
p.138	in figure 11.7 caption, line 4, for "(A)" read "(a)", and line 5 should read "(b) as for (a) but..."
p.146	reference 59, for "in Press" read "*149* (1978) 297"
p.154	in lines 4, 12 and 13, for "t_o" read "t_m"
p.160	line 2 from foot should read "with $V_R = 2000$ mm^3, ..."
p.198	appendix table: refractive index of 1,1,2-trichlorotrifluorethane is 1.357, not 1.250 boiling point of hexane is 69.0°C, not 86.2°C

INDEX

Abbreviations for types of chromatography:
LSAC Liquid-solid adsorption
LLPC Liquid-liquid partition
IEC Ion-exchange
EC Exclusion
IPC Ion-pair
RPBC Reversed-phase bonded
HPLC High-performance liquid

References to tables or figures are indicated by the letters T, F, respectively.

Date Du
MY 9 '89
DE 03 '90
DE 15 '92
FE 23 '93
MR 17 '93
MR 29 '93
BRODART, INC.
Cat. No. 23 233